Farming Book Series

FARM CROPS

GRAHAM BOATFIELD

Illustrations by Keith Pilling

FARMING PRESS LIMITED

WHARFEDALE ROAD IPSWICH SUFFOLK ENGLAND

First published 1979
Reprinted 1980
Second edition 1983

ISBN 0 85236 129 7

Set in eleven point Times and printed in Great Britain on Longbow Cartridge paper by Spottiswoode Ballantyne Ltd, Colchester, for Farming Press Limited.

FARM CROPS

Contents

APPENDICES

Illustrations

Introduction

THIS BOOK is designed to give a simple outline of farm crops and crop husbandry, as found on British farms today, along with the basic knowledge and the scientific principles underlying the growing of those crops and the use of the land to best advantage.

The course of study set out here needs positive work by the student, based on farm observations and simple studies carried out locally. A book of this type gives a very general picture—local information is vital to complete that picture. Crop behaviour and crop yields vary according to soil types, climate and systems of farming. National averages and general comments mean little if they are not related to local conditions.

This book is in six sections. *The Soil* covers the composition of soil and the nature of all the things found in any farm soil, with an outline of the problems presented to anyone who has to handle soil and use it for any farming or growing process. *The Plant* covers plant types, structure and growth, and the problems found in growing commercial crops. The section on *Farming* discusses and relates all those factors of soil and plant which can be controlled and influenced by the farmer, the farm worker and the scientist.

Dealing directly with crops, there are three more sections. *Combinable Crops* deals with cereals and many of the break crops now commonly grown. *Root and Forage Crops* covers the main cash crops and farm-consumed crops grown for human, industrial and livestock use. *Grassland* aims to simplify a complex subject and deal briefly with our most important crop.

No information about crop varieties is given in this book; there would be no point in doing so. Varieties change, even from year to year, and the claims of seedsmen are not always confirmed in large scale practice. Everyone concerned with farming should get to know the publications of the National Institute of Agricultural Botany (NIAB) of Huntingdon Road, Cambridge, in which suitable varieties of many crops are assessed and recommended. There is a special scheme of membership of NIAB for students, which provides useful information at low cost.

The figures given for fertiliser applications are average figures; in practice they should be modified according to soil conditions, local needs and cropping systems. There is plenty of good advice available to farmers and growers, from the Ministry of Agriculture (ADAS) and from commercial firms, both on fertilisers, and on the use of sprays and other chemicals.

By using a crop calendar form based on that shown on page 140 of this book particular crops can be studied. To make a thorough study of farming, a proper farm diary is essential.

Farm Crops is one of a set of four books specially prepared to meet the needs of students, those new to farming, and anyone who wants to understand modern farming practice. The other books in the series are *Farm Machinery* and *Farm Workshop* both by Brian Bell and *Farm Livestock* by Graham Boatfield. Study of these books in combination will give a sound basis of knowledge of mechanised arable and livestock farming.

The material presented in these books covers the requirements of students taking courses which lead to examinations at Phase I and Phase II level of the City and Guilds of London and the regional examining bodies. They will be of value for school work in land-based subjects and rural and environmental studies, and for the private study of anyone with an interest in farming.

Any reference to farm safety or other regulations are given here purely as a guide. The reader should refer to the relevant leaflets and other publications to find the exact requirements of the current regulations.

Figures used in these books are stated in both metric and British terms in common use, for the convenience of readers. Metric figures and dimensions are adjusted conversions and not necessarily exact.

My grateful thanks are due to my professional colleagues, Mr Stewart Cairns formerly of ADAS and Mr John Gibson of the Lincolnshire College of Agriculture who have helped with the processing and preparation of material used in this book; and to Mr Philip Wood for his help and advice.

Chapter 1

The Soil

THE SOIL is the one basic farming material. Everything else depends on it and on its productivity. To handle soil properly, and to produce the most and the best from it, we must understand it fully, by theory and by practice.

SOIL MATERIALS

A soil is not just a haphazard collection of material. It is a living thing. It must be living to grow plants, and to make all the chemical changes needed in the substances that are added to the land.

Soil Layers

Soils are formed in layers, and the way in which the layers are arranged depends partly on natural conditions and partly on management and cultivation. The main layers of the soil are these:

The Rock Below. This is the material that was there in the first place, from which the soil was made. It may be any of the main rock materials (called rocks, but some of them are not rock-like in the normal sense)—sandstone, limestone or other hard rocks; chalk, sand, silt, clay or peat.

The Subsoil. This is the rock partly broken down and altered in some way by action from the top of the soil, with fine material which has worked its way down from the top. There is usually no organic matter in the subsoil, with the exception of the deep peaty soils, and it is like dead soil. If subsoil is brought to the top by deep cultivation, it is 'raw' and it may take some time to weather down and work like a proper soil.

The topsoil in cultivated land varies from about 8 cm (3 in) to 30 cm (12 in) deep, according to the nature of the subsoil, and the depth of cultivation. It is darker because it contains organic matter, is freer working, and is more manageable than the subsoil. It is a real living soil; plant growth and crop production depend on the condition of this shallow layer of soil.

There is only one way to see the whole soil as it really is, in layers like a sandwich: look at a *soil profile*, which is a cross section of the layers of the soil. You can see this in a pit, the side of a quarry, or sometimes in a road embankment.

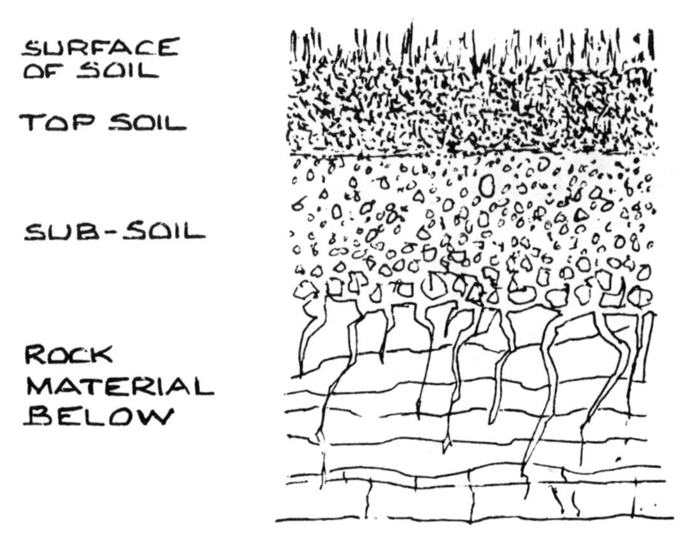

Fig. 1. Layers of the soil (a soil profile).

Soil is a Mixture

It is a mixture of quite a number of different things:

Living Things: very small plants such as bacteria and fungi, and small animals such as insects and earthworms;

Soil Air and Soil Water: the soil air contains more carbon dioxide than ordinary air, and the water contains plant foods;

Organic Matter is of two main types. The raw organic material added to the surface of the soil, sometimes called *fibre*. The organic matter which has been broken down in the soil to a dark spongy substance, known as *humus*;

Mineral Matter comes from the rock material below, and may be any one or more of the following: rocks, stones, gravel, sand, silt, clay.

Most of our soils are a mixture of much mineral matter with a little organic matter. The main difference between these grades of mineral matter in the soil is size. Rocks, stones and gravel are fairly large, and clay very small with silt and sand in between. This table shows the sizes of particles of these soil materials:

Soil Material	*Diameter in Millimetres*
Stones and gravel	Above 2
Coarse sand	2·0–0·2
Fine sand	0·2–0·02
Silt	0·02–0·002
Clay	Below 0·002

Fig. 2. Size of soil particles (magnified 40 times)

The Special Nature of Clay. The particles of clay are very small indeed. According to the way the land is treated, they either all go into a solid mass (like the bottom of a pond, or the surface of clay soil near a gateway which is called 'puddled' or 'poached') or separate out into crumbs; this is known as 'flocculation'.

A mass of clay is just like a sponge, and has so much space between its tiny particles that it can hold a tremendous amount of water.

How Soils were Made

Most soil materials have come from the rock below, the mineral part of the soil which has been weathered and broken down, and mixed with organic matter made from plants which have lived in the soil.

It is useful to know just how the mineral part of the soil has been made by the weathering of the rocks below.

Weathering of Rocks

A number of different forces act very slowly on the rocks which causes them to *weather* and break down. These forces include:

Frost which freezes water in cracks in the rocks, and breaks them up by expanding;

Water which wears away the sides and bottoms of streams, and the rocks under waterfalls;

Heat and Cold which shrinks and swells the surface layers of the rocks, and makes them crack and crumble;

Wind which blows sand against larger rock surfaces and breaks them down, and which blows away sand and dust and leaves it elsewhere;

Chemicals, in the rain and the waters of streams, which help to dissolve away some of the rocks;

Ice: the movement of large masses of ice, such as glaciers, over the surface of the rocks breaks them down, carries bits and pieces away and leaves them elsewhere;

Plants: The roots grow into the crevices and holes among the rocks and slowly help to break them down.

Fig. 3. Pore spaces of soils.

SOIL TYPES

In a practical farming way, the two main types of soil are light and heavy, but there are mixed or medium soils which combine some of the qualities of the two extreme types.

Light Soils are easy to work, need less power to cultivate, can be worked at most times of the year and do not hold water so much. They include the sands and gravels, chalky soils and some of the peaty soils.

Heavy Soils are more difficult to work, need much more power to cultivate, can only be worked at certain times when they are in the right condition, and hold water. They are usually more productive and grow heavier crops. Heavy soils usually contain plenty of clay.

This table shows how light and heavy soils compare:

	Light Soils	*Heavy Soils*
Movement of water	Passes through easily and quickly	Passes through slowly; usually need drainage
Holding water	Do not hold much	Hold a lot of water
Ease of working	Need little power	Need much more power
Time of working	Can be worked at any time	Only work them at certain times, or there is trouble
Time of grazing with livestock	Any time	Not too early nor too late in the year
Start of plant growth	Early in the spring	Late
Ploughing	Autumn or spring	Autumn only, to allow the frost to work on the soil
Natural fertility	Low	High
Size of particles	Large	Very small

How to Tell Soil Type

Always be prepared to handle soil; it will tell you much more than just looking at it or even walking over it. You can make sure what type a soil is by a few simple hand tests, but either take moist soil or moisten it.

1. If a sample of soil feels *gritty*, it is a *sand*.
2. If it feels gritty and makes your fingers dirty, it is a *loam*.
3. If it feels *silky*, but you cannot polish it when damp between your fingers, it is *silt*.
4. If you can polish it between your fingers, it is *clay*.

(The loams, silts and clays all cling together.)

The mineral matter in a soil may be rock, gravel, sand, silt or clay. According to how much there is of these different materials, so the soils can be named.

The Main Soil Types

Clay Soils contain more than 30 per cent of clay. They are sticky when wet, hard when dry. Organic matter takes a long time to break down in these soils, which are often fertile, though sometimes low in phosphates (P). Clay soils are late in the spring, slow to warm up. They cannot be worked when wet, and should be cultivated only at certain times. Crops which do well on heavy clays include wheat, beans and grass.

Sandy Soils contain no more than 5 per cent of clay. They are open, free draining, quick drying. They are 'hungry soils', quickly using up organic matter and plant foods; low in fertility, with not much plant food, and often sour and short of lime. Crops which do well on sandy soils include barley, rye, root crops and vegetables.

Loams contain a balanced mixture of light and heavy soil materials (usually sand and clay) and are the best all-round soils. They are naturally fertile, free working, easily drained, and will stand up to very dry conditions. Soils containing a lot of silt are often like good fertile loams.

'Limey' Soils (sometimes called calcareous soils) contain a large proportion of calcium carbonate, either in the form of chalk or limestone, according to the rock below. Chalky soils dry out easily and bake hard in summer; in the winter, frost raises the soil and can lift plants out of the ground. These are usually hungry soils, needing organic matter and often low in phosphates (P) and potash (K). Some of the limestone soils are good loams containing lumps of limestone from the subsoil. Crops which do well on 'limey' soil include barley, grass and clover, seed crops and sugar beet where the soil is deep enough.

Peaty Soils contain 50 per cent or more of organic matter. When well drained these are rich and fertile soils, as found in the Fens. If badly drained, they are sour and difficult to crop. Often high in nitrogen (N), they are commonly short of potash (K) and lime. Heavy crops are produced, but the quality may be low. Crops which do well in the Fens include wheat, potatoes, celery and flower bulbs.

Stony Soils are found in some places where large lumps of the subsoil material are present in the top soil, and have not weathered down. Some of these soils are difficult to work, and cause heavy wear on cultivating implements. Examples of poor stony soils are the shallow stony soils found in some hilly districts. Good soils of this type are the limestone soils of the Cotswolds and Lincolnshire.

SOIL STRUCTURE

We have already seen that soil is not just a mixture of different substances, but is arranged in a definite way, usually in layers. In most soils, the smaller particles cling together, and often larger lumps separate out to form a soil which can be farmed and cultivated properly. This arrangement in a soil is known as soil structure.

Poor Soil Structure

Think for a moment of two types of soil material without structure. A heap of pure sand is the same right through; it handles easily, but does not hang together; it holds very little water; it just does not seem like a living soil. A large lump of wet clay is also the same right through; it is solid, and although it holds water, it is not easy for water to pass through; there is little room for air in it; plants find it difficult to grow in it and roots to develop. Both these are extreme examples of soils without structure, which are difficult or impossible to farm. You can find examples of these extreme types on a sea-shore or at the bottom of a pond. Clay which has been trodden too much by animals (as at the gateway into a field) or worked and smeared together by implements under wet conditions, goes very solid and is then known as a *poached* soil.

A layer under the soil which has been smeared by implements becomes a *pan*—a solid layer. The same effect can be produced by the action of iron salts in the soil—a rock-like layer.

Good Soil Structure

Think of a good rich loam soil; handle some when you get the chance. This breaks to pieces when handled, yet still sticks together if it is pressed into a lump. Water can pass through it easily; there is plenty of room for air it it; plants grow well in it, and the roots can spread without difficulty. From the farmer's point of view, it is a soil that can be cultivated and farmed easily. We say it is a soil with a *good crumb structure*, a *good tilth* or that it is a *kind soil.* The topsoil breaks into crumbs, so that air and water can move between them, yet each crumb clings together and holds some supplies of water and plant food.

Why Soil Structure is Needed

These are the main reasons for good structure or condition in the soil, so that:

1. Seeds can germinate and get the conditions they need for the young plant to grow properly.
2. Roots can spread through the soil, and the small root hairs take in water and plant foods.
3. Water and air can move through the topsoil.
4. Water and fertilisers are taken into the surface of the soil, and the water does not run off the top.
5. The surface of the soil does not become hard and crusty, which prevents plants growing properly.

What Makes Soil Structure

Two main things give structure to soil: clay and organic matter. It may be either one or both of these but it is important that the clay is in the right condition. Both clay and organic matter make soil cling together and hold water. Roots of plants also give some structure to soil; you can see the effect of roots in a sandy soil by pulling out a plant and seeing how the grains of sand cling to the root.

How Soil Structure is Improved

Heavy soils: The clay needs to be weathered and cultivated properly; both are important.

Weathering includes the following:

1. Frost in the winter splits the clay into smaller lumps.
2. Wetting and drying, one after the other, breaks the clay down into crumbs.

It is important to cultivate clay soils at the right time. This needs knowledge and experience. For example, the right time to break down clods is when they are just 'between wet and dry', either drying out or recently wetted by rain.

Light soils: usually the only thing that can be done is to add organic matter. These soils sometimes form a crust on the top in the spring, and this should be broken down by harrowing or other cultivations so that air and water can get into the topsoil. If a *pan* (hard rocky layer) has formed under the topsoil, the soil can only be improved by cracking this pan by deep cultivation.

SOIL ORGANIC MATTER

This is part of the soil which has been made from living things, mostly plants. It is not in itself a living substance, but because it was once alive it is *organic* and thus different from the main part of the soil which is dead *mineral* material. Soil organic matter is important for its effect on the structure of the soil (see page 16) and because it provides a food for the living things in the soil (see page 20).

Types of Organic Matter in the Soil

Although organic matter is made of many sorts of different things, there are only two main types of it in the soil, one always being turned into the other.

Raw Organic Matter which is the fresh material added to the top of the soil, worked in and gradually worked down through the topsoil. It looks like what it is—fresh or partly decayed and broken plant remains. As it is broken down it turns into:

Humus. This is the completely broken down organic matter. It does not look at all like the stuff it is made from, but is dark, crumbly, sometimes powdery, spongy material and spread through the soil.

How the Soil is Supplied with Organic Matter

Both natural and cultivated soils are supplied with organic matter from the surface, in the ways listed below. A few special soils, such as the Fens, the moors and peat bogs, have great deposits of organic matter (peat) like a rock material beneath them, and this is often very old. Farming and cultivation bring some of this material gradually into the topsoil. Normally, organic matter comes into the topsoil from the following:

1. *Plant Remains*. Leaves of trees and plants, potato tops, cereal chaff, straw and stubble, sugar-beet tops, special crops grown for 'green manuring'.
2. *Roots* of all crops and weeds.
3. *Farmyard Manure* and other organic manures such as dung from grazing animals, litter from poultry houses, sawdust and wood shavings.
4. *Grass*, either leys or permanent grass, when ploughed in adds more organic matter to the soil than any other method.

Today, farmyard manure is used less than in the past. It is not always farming practice to plough up grassland, and so it is often found that most organic matter comes from roots and plant remains.

How the Soil Loses Organic Matter

Some of the raw organic matter is broken down into humus, but not all of it. Some of it is burned up (oxidized) by bacteria, and by this process it is turned back into the substances which the plant used right at the beginning to make its own substance—water and carbon dioxide and plant foods. How these bacteria break down organic matter is discussed in the next chapter.

The 'hungrier soils', usually the lighter sands, break down organic matter quicker than the heavier colder soils.

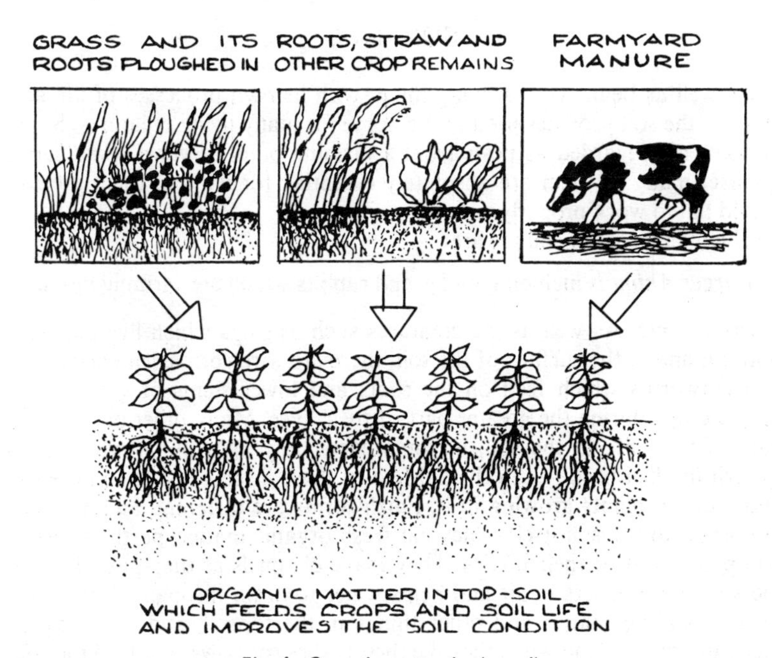

Fig. 4. Organic matter in the soil.

How Organic Matter Affects the Soil

Organic matter has a good effect on any soil. It is needed most in the extreme types of soil—the sands and the clays—to make them less difficult to farm. The main effects of organic matter are:

1. It helps the soil to hold water, which is needed by the crops.
2. It helps light sandy soils to cling together, and gives them some structure.
3. It opens up heavy clay soils and improves their structure.
4. It feeds the living things in the soil.
5. It supplies plant foods, which are slowly released to the plants as the organic matter is broken down.
6. Because humus is dark brown, it makes soil darker and so it warms up more quickly than the lighter soils.

How Organic Matter Affects Crops

A soil containing plenty of organic matter handles better, makes a better tilth, and provides the best conditions for seeds to germinate and for the young plants to grow. As it gives up plant foods slowly, this often helps the plant throughout the growing season.

SOIL LIFE

As well as being itself a living thing, with its own processes of life and change, the soil provides food and a home for many living creatures. Some of these are regarded as pests from a farming point of view; some cause disease; many of them are absolutely essential, for without them the soil would fail to work properly.

The main types of soil life are:

Larger Animals including moles and rabbits which are farming pests.

Insects, etc. As well as the creatures such as slugs which live and feed both on and in the surface of the soil, there are wireworms, leather-jackets and cutworms which feed on the roots and lower stems of grasses and arable crops during the greater part of their lives. Many other pests spend part of their life in the soil, in some cases only as a pupa: wheat bulb fly, frit fly, flea beetle and others. Most of these insects just live in the soil and feed on plants growing in it. Earthworms are different, and are very important to the soil itself. They eat fresh organic matter, partly digest it, and pass it out of their bodies. They make channels as they pass through the soil, and this lets in air and helps water to drain through. They churn up the organic matter and other things in the topsoil, and thus help to cultivate the soil, and when they die their bodies rot down and feed the soil. A soil without earthworms is a dead soil.

Micro-organisms are all the very small living things which are found in the soil, as in so many other places above ground.

(a) *Viruses* are living chemicals existing in the soil and in some cases cause disease.

(b) *Fungi;* many of them live in the soil and have their work to do in the decay of organic matter. Larger examples are the many different mushrooms and toadstools which feed on organic matter. Some diseases are caused by the soil-borne fungi which live below ground for part of the time and which infect crops and other plants (see page 43).

(c) *Bacteria* are very small one-celled plants which grow quickly and multiply by splitting in two. To grow properly they need three things: moisture, warmth and the right food. In some cases they need lime to neutralize the acids they produce. There are thousands of different types of bacteria, and most of them do just one job—some cause diseases, others make chemical changes, etc. Some grow best when air is present (aerobic bacteria) and others only grow without air, as in very wet conditions (anaerobic).

It is the bacteria which are responsible for most of the soil life that is necessary for soil fertility. These are examples of the work done by bacteria in the soil:

1. Breaking down some organic matter into simpler forms, including humus.
2. Burning up (oxidizing) organic matter so that it is turned into water and gas.
3. Changing certain chemicals containing nitrogen (N) into other forms:
   Ammonium compounds to nitrites (Nitrosomonas bacteria)
   Nitrites to nitrates (Nitrobacter bacteria)
4. Taking nitrogen (N) from the air and making it into organic matter in the soil. (Azotobacter bacteria)
5. Special bacteria (called Rhizobium) live in the soil and pass into the root nodules (small lumps) which form on the roots of all plants of the legume family (see page 30). There they live with the plant, which gives them the foods they need and in return they feed the plant with nitrogen which they take from the soil air and make into a form which the plant can use.

For the common legume crops (beans, peas and clovers), these bacteria are found normally in our soils, but for lucerne, a special strain of bacteria is needed which is not in British soils. For this reason lucerne seed has to be inoculated with these bacteria, supplied by seedsmen in a culture which comes with the seed. We should have to do the same sort of thing if an unusual legume crop was grown anywhere for the first time.

SOIL AIR AND WATER

The spaces between the solid particles that make up a soil are known as *pores*. These pore spaces in any soil are full of soil air, soil water or both. Water forces air out of soil and this happens in a soil which is too wet. For the best possible conditions, for the health of the soil and the proper growth of plants, both are needed. A soil with a good crumb structure is in the best condition for soil air and soil water to move and to their work properly.

Soil Air

The air in the pore spaces of the soil is different from the air above the ground. It is wetter (contains more water vapour), contains more carbon dioxide and nitrogen, and less oxygen.

Soil Water

No soil is completely dry, there is always some water inside the particles, no matter how dry it seems. Some of this water can be used by the roots of plants; some is held too firmly inside the soil particles and plants cannot get it.

How the Soil Holds Water

Below a certain level in most soils, the land is full of water. The top of this waterlogged soil is known as the *water table.* It varies according to time of year, amount of rainfall, type of soil and how the soil is drained. If the water table comes too near the surface, most of the soil is waterlogged. In waterlogged soil, the pore spaces are full of water, and there is no air. Roots do not grow in water. Arable crops need a low water table, so their roots can develop, which produces better yields. Grassland can grow well if the water table is higher. For this reason some marshland will grow grass well but not arable crops.

Water is also found as a thin film around the actual particles of soil except in very dry conditions. The roots of plants can use this water. Water rises from the water table through the soil in the same way that liquids rise through the wick of a lamp, blotting paper or a very narrow tube (this is known as capillary action). The finer the particles of soil the higher the water will rise in this way.

How the Soil Loses Water

The water in the soil is lost in several ways. This is a much more serious problem in dry districts where water is short in any case.

1. *Drainage:* water passing through the soil and out through drains or through the sub-soil.
2. *Evaporation:* water at the surface of the soil turns into water vapour and goes into the air. This happens much more quickly in hot conditions, and where there are dry winds blowing over the top of the soil. In very dry conditions rain water is evaporated from the soil almost as soon as it falls.
3. *Through Plants:* every growing plant is taking in water at the roots and passing it up through its stem to the leaves, where it is given off to the air as water vapour. This is known as *transpiration* (see page 34). A great deal of water is lost from the soil in this way.

CHEMICALS IN THE SOIL

We are only interested in the chemistry of the soil that is concerned with feeding plants, growing them properly, keeping them healthy and keeping the soil in good condition. Most of the mineral matter of the soil is not concerned with feeding plants, it merely supports them. We are concerned with the chemical substances dissolved in the soil water or held in one form or another in the humus, the organic matter and the clay.

The Effect of Farming

In most soils there is a good supply of all the plant foods. This would supply all the needs of the plants that would grow naturally if the land were left under natural conditions. As these plants die and rot down in the soil, the plant foods would be freed in the soil and used again.

When we come to farm the soil, we grow crops which are very much heavier than those grown under natural conditions. As we cut these crops and take most of them away, large amounts of plant foods are taken away and most of them are not returned to the soil.

Farm crops are quick growing, and need their plant foods during a very short time of growth. Plant foods found naturally in the soil are only released slowly by chemical change and by the action of bacteria. For this reason, quick-growing crops cannot get all their plant foods naturally. We have to use fertilisers to feed the crops we grow.

Plant Foods in the Soil

All the main plant foods (see page 34) are found in a normal soil. Some of each plant food is there in a form the plant can use straight away. This is known as *available* plant food (such as available nitrogen). Some more of it is in a form which the plant cannot use at the moment which is said to be *unavailable*, but to become available it may need to be changed chemically in some way or broken down by bacteria or in some other way in the soil.

Nitrogen which is naturally in the soil is mostly in the form of humus and organic matter which the plant cannot use until it has been broken down into nitrates. This happens as follows, by the action of different types of bacteria:

Organic Nitrogen → Ammonia → Nitrites → Nitrates

Phosphorus in the soil is nearly all 'locked up' as far as the plant is concerned, as it is in an unavailable form. This is released to the plant very slowly as organic matter is broken down. When fertilisers are applied some of the phosphorus is used by the plant at once but very soon most of it becomes locked up and only released slowly afterwards. The plant takes in this food in the form of *phosphate*.

Potassium is held in the soil in organic matter and in clay, and is released to plants steadily.

Trace Elements, the other plant foods which are needed in very small quantities, are found naturally in most soils. There are a few cases where

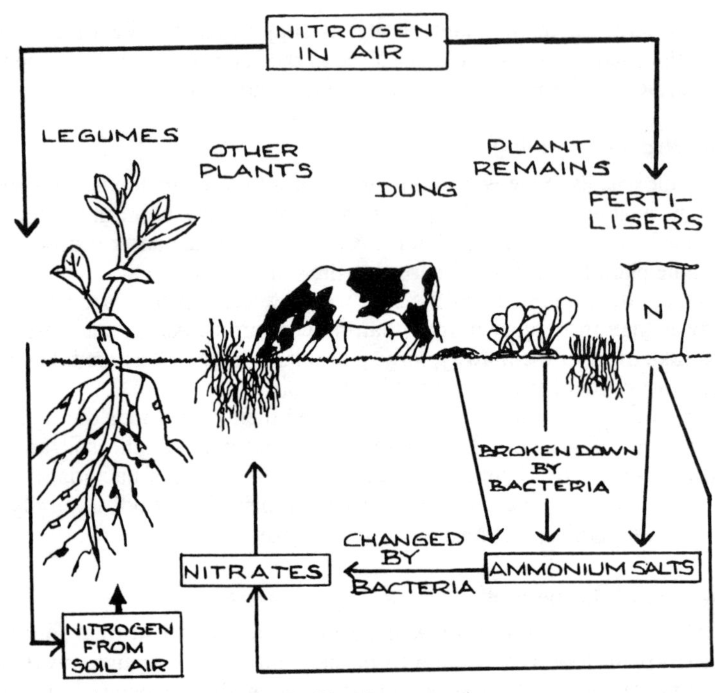

Fig. 5. The nitrogen cycle.

one particular plant food (such as boron, copper or cobalt) is short and this *deficiency* can cause trouble to crops, livestock or both. As heavier crops are grown and these trace elements used up more quickly, more shortages are found under normal farming conditions.

Lime in the Soil

Lime, which consists of various forms of the element *calcium*, is needed by the soil for various purposes, the chief of which are:

(a) to prevent the soil becoming acid or 'sour', by neutralizing acids formed in the soil by the action of bacteria and the breakdown of organic matter;

(b) as a plant food for all crops, particularly for plants of the legume and cabbage families;

(c) to encourage soil life (bacteria etc.).

There is plenty of lime in some soils, those with a subsoil of limestone or chalk (which are both forms of calcium carbonate) or containing small pieces of chalk, such as the boulder clays. These soils do not normally need to have any lime added. Many other soils are short of lime, both in the topsoil and the subsoil and if lime is not added this can lead to poor cropping and other troubles—the sandy soils are worst in this way.

Loss of Lime from the Soil

Lime is removed from the soil in two ways:

(a) by water passing through the soil and washing away (leaching) the lime which it dissolves in small amounts;

(b) in crops and grass. Where these crops are grazed, lime is removed by livestock which need lime for their own bodies (bones, teeth and milk production).

With normal cropping, the equivalent of 50–200 kg (1–4 cwt) of lime is lost from an acre of land each year. An *acid soil* is a soil which is naturally low in lime or from which the lime has been removed.

THINGS TO DO

1. Visit a local sandpit or quarry to see a soil profile. In the same way, study your local soil types by looking at holes, pits or embankments.
2. Walk over soils of different types, and handle the soil. Look at soil samples side by side to compare them. Compare soil with plenty of fibre in it (like grassland soil) with ordinary arable soil.
3. Dig holes in the soil, look at the different levels of the soil, look for 'pans' under the soil.
4. Use a soil auger to take samples from different layers of the soil, test these samples for acidity.
5. Look at methods of draining soil and methods of applying water to crops.

QUESTIONS

1. What are the three main layers in the soil?
2. What is the special quality of clay in any soil? How does this affect the farming use of the soil?
3. What can be done in practice to improve the structure of soils?
4. How is organic matter added to the soil, and how is it lost from the soil?
5. Practically, how can you tell one soil type from another?

Chapter 2

The Plant

A STUDY of the plant, its structure, development, growth and troubles is necessary to understand the scientific growing of crops and the proper use of the soil to grow the best possible crops.

PLANT STRUCTURE

There are two main parts of the plant, the root system below the ground and the stem and leaves which are above the ground. This is putting it very simply because there are some cases where roots are found above ground and stems grow below. Looking at the plant in more detail, these are the parts and the jobs they have to do:

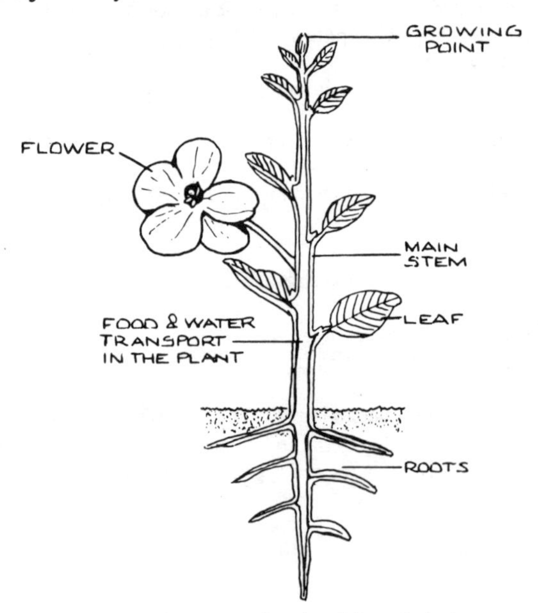

Fig. 6. Structure of a plant (dicotyledon).

The Root fixes the plant in the soil, takes in food and water, and in some cases stores food. The roots come from the base of the stem and divide many times, getting smaller and smaller. Smallest of all are the *root hairs* which are found along the smaller roots; these take in water and plant foods from the soil. Plants have two different types of root system. Some, including root crops, peas and clovers, have a main *tap root* which goes straight down into the soil, with other roots branching from this; sometimes this tap root is very thick and tough. Other plants, including the grasses and cereals, have *fibrous roots* like a tuft (see below). The roots push their way through the soil as the plant grows; spreading wide and deep with many of our crops but will not grow into waterlogged soil.

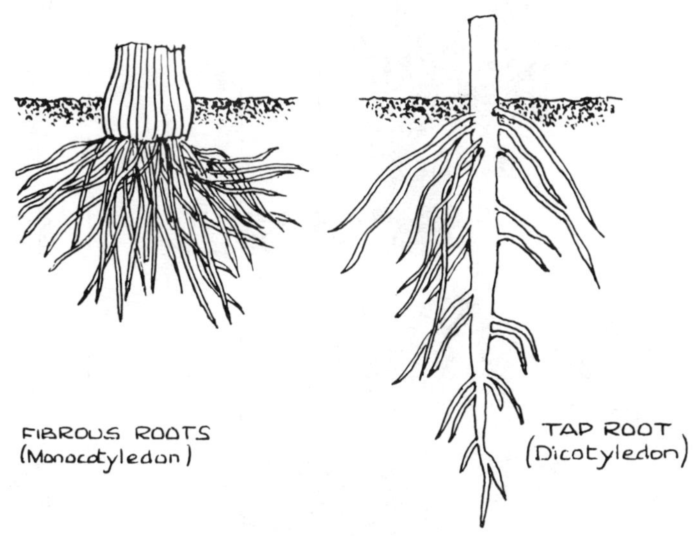

Fig. 7. Types of root growth.

The Stem supports the leaves and flowers which grow out of it, acts as a channel between the roots and the leaves, taking water and foods from one to the other, and sometimes is used for storage. The stem gets longer as the plant grows, and contain various tubes through which the sap of the plant (water containing foods etc.) moves.

The Leaf is green, thin and arranged on the stem so that it gets as much light and air as possible. Inside the leaf are veins which strengthen it and carry water and foods. There is a green colouring, *chlorophyll*, inside the leaf which is important in the making of food for the plant. In the skin of the leaf are a number of little holes, each one a *stoma*, and water is given off through these at certain times and gases pass in and out. Each leaf is a factory for the plant where food is made; it also provides energy for the

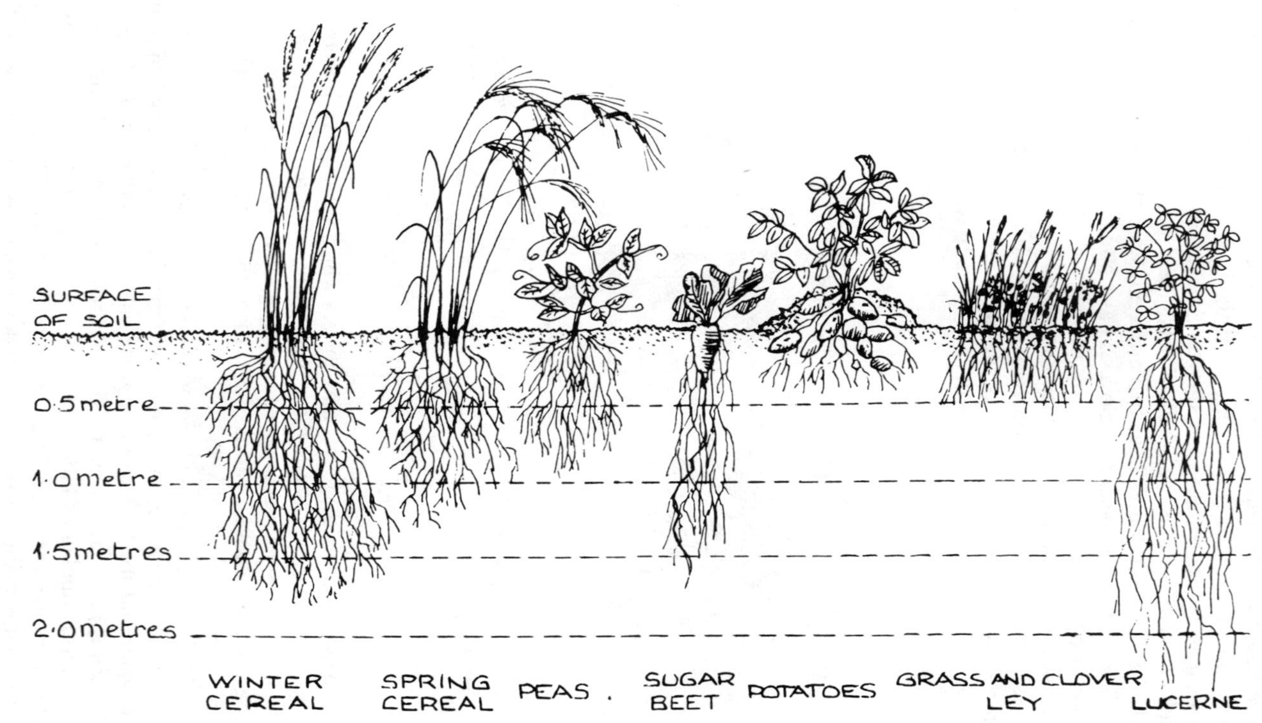

Fig. 8. Roots of farm crops.

plant by the process of respiration which uses up some of the food (see page 34).

The Flower only appears at certain times in the life of the plant. The stamens, the male part of the flower, produce pollen which fertilises the female part of the flower, and this leads to the formation of seed which ripens on the plant.

TYPES OF PLANT

There are thousands of different plants which grow naturally; only a few of them are important in farming.

The Use of Plants for Farming

The average farmer sorts out plants into three categories; crops, weeds and the rest.

The crop plants can be sorted into main groups according to their type and the reason why they are grown:

Cereal crops, usually known as corn, include wheat, barley, oats, rye and maize, all grown for their grain.

Other combinable crops, which can be harvested in a similar way to cereals, include peas and beans (which are sometimes known as pulse crops), oil seed rape, linseed and a large variety of seed crops.

Root crops, grown for their roots which are sold or fed to livestock, include potatoes, sugar beet, mangels, swedes, turnips, carrots and other root vegetables.

Forage crops, grown for their leaves and stems—sometimes fed direct to livestock, sometimes harvested and processed first—include kale, cabbage, forage maize, forage rape and radish, mustard, etc. This group also includes grasses, clovers and lucerne, which are sometimes known as *herbage crops*.

Length of Life

Crops, weeds and all other plants can be divided into:

Perennials live for more than two years (often a long time) and grow each year. Some are *woody*, such as trees and shrubs, and others are *herbaceous*, softer and less persistent, such as the perennial grasses, white clover, thistles and docks.

Biennials live for two years, storing up a food supply in the first year; after resting for the winter, they flower and produce seed in the second year. This group includes sugar beet, swedes, most of the other root crops, and a few weeds like ragwort.

Annuals live for one season only, producing seed and then dying. Cereals, beans, peas, mustard and most of the common weeds are examples.

Groups of Plants

There are two main groups of plants, according to their number of *cotyledons* (the seed-leaves which are found inside the seed).

Monocotyledons include grasses, cereals, rushes and onions. Generally they have no tap-roots, a number of equal shoots and their leaves are long and narrow with parallel veins.

Dicotyledons include all the other farm crops and most of the weeds. They have tap-roots, usually a main stem with branches coming out of it, and their leaves are broader and shorter, with branching veins.

For practical purposes these groups are different in the way they grow and in the way they react to certain weedkillers.

Families of Plants

There are about 200 families of plants but only a few concern us. Crops belong to the following families which also contain weeds. Knowing these family divisions may be of some use, as pests and diseases sometimes attack or live on several plants of the same family:

Graminae: Grasses, cereals, and grass weeds.
Leguminosae: Peas, beans, clovers, lucerne.
Brassicae: Cabbage, kale, turnips, swedes, rape, mustard, charlock.
Chenopodiaceae: Sugar beet, mangel, fat hen.
Solanaceae: Potato, tomato, nightshades.
Umbelliferae: Carrots, parsnip.

PLANT GROWTH

Unlike an adult animal, a plant may be growing during nearly all its life.

How the Plant Starts

There are two main ways in which plants reproduce themselves, described at the end of this chapter. Whichever way it is, the plant starts from something small: a seed, tuber, bulb or cutting.

Germination of the seed starts when the seed is ready to grow and the conditions are right for it. The three things necessary for this are:

Air, *Water* and *Warmth*.

The seed is a living thing: it must breathe and when growth starts it needs more oxygen, so air is needed in the soil. A seed will not germinate in a waterlogged soil; it would die and rot. Water is needed for growth and for the chemical changes that take place inside the seed. Until the temperature is right, the seed will not germinate; the right temperature lies between 10°C and 45°C (50°F and 110°F).

Different types of plants have different methods of germination, but the general idea is the same with all of them. The food stored in the seed is

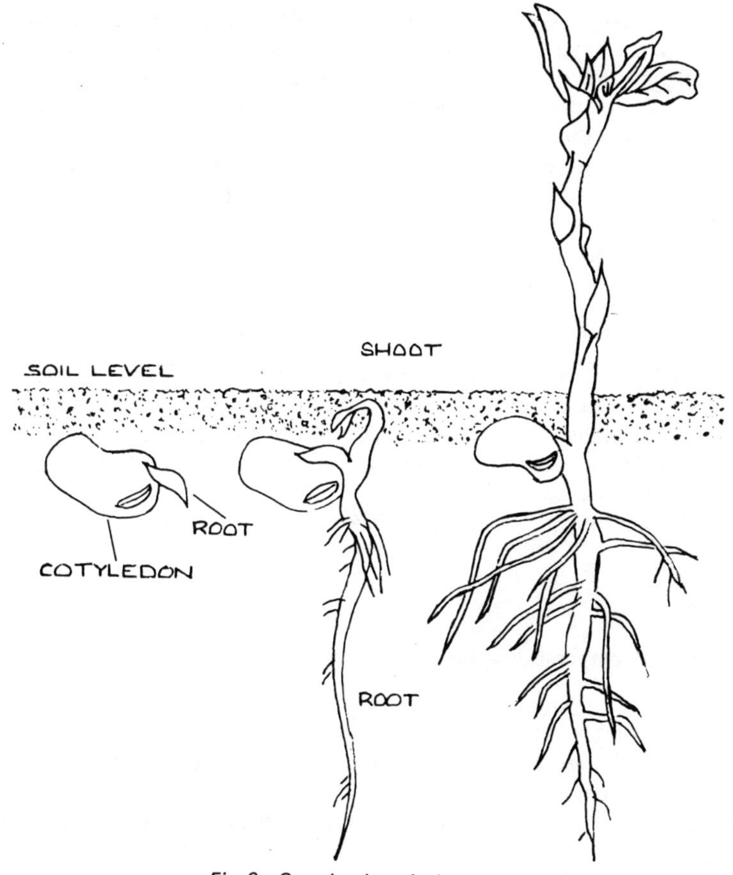

Fig. 9. Germination of a bean seed.

used to make the young shoots and roots grow until the plant can get its own food and water; just the same as the egg and the young chick.

The starchy food supplies are changed into sugars which can be used to power the new plant which the seed is making (this process is used commercially in malting by which barley seed is made to start growing and is then stopped after a few days (see page 78).

The young shoot (the plumule) and the young root (the radicle) swell and burst out of the skin of the seed. The shoot presses through to the surface and then develops into stem and leaves. The *Cotyledons* may stay below ground, as in wheat or beans, or they may appear above ground as two first leaves, as in cabbage or swede.

The young root grows down and produces more first roots till the seedling can start to feed itself and growth is very speedy.

In these early stages, there is a difference between the groups of plants. *Dicotyledons* have a bent shoot (see fig. 9) which straightens above ground, and they produce a tap-root. *Monocotyledons* press straight up and appear as a first leaf; they develop another root system at the bottom of the stem and the first roots die away.

First Growth from tubers, bulbs etc. is different from germination. The plant 'sprouts' and produces shoots and roots which keep on growing and feeding. Sometimes the original piece dies away and rots in the soil, as with potatoes, or it may form part of the new plant, as with couch grass.

How the Plant Keeps on Growing

A plant grows by adding the food it makes in its leaves and the water it takes in through the roots to its own body. Most growth takes place in parts of the plant which develop very much faster than the rest. In roots this is just behind the tip; in shoots it is in the upper parts; and in leaves it is the parts nearest to the stem.

To make proper growth the plant needs *all* the following:

Plant foods.
Water.
Air.
The right temperature—neither too hot nor too cold.
Light.

The *growing point* of a plant is the place from which it makes its growth. In *Dicotyledon* plants it is near the top of the stem; if it is damaged or removed, the plant will branch out sideways. In *Monocotyledon* plants the growing point is at ground level; for this reason grasses can be grazed off many times and will still keep on growing, also producing side shoots (*tillers*) which branch out from ground level.

Reproduction

Many plants—annuals and biennials—only live until they have flowered and produced seed. Perennials also produce seed, but go on living after this although seeding may weaken them as with the grasses.

Seedling. Most of our crops are reproduced by seed. The plants flower and the female part is fertilised by the male pollen. (With most of our crops the male and female parts are both present in the same flower.) The seed develops on the parent plant and when fully developed it is separated by harvesting and some form of processing. Seed may produce a plant which is just like the parents but crossbreeding between different plants (and perhaps between different varieties or types) may produce plants which are better or worse than the parents.

Vegetative Reproduction is found in some plants which may also produce seed. It means taking a piece of the original plant and using it to make a new plant, or forming new plants which are attached to the parent, both parent and new plant being exactly the same. Common farm examples are:

Tubers: Swollen underground stems full of food reserves; potatoes.
Rhizomes: Underground creeping stems; couch grass.
Stolons: Above-ground creeping stems; wild white clover.
Runners: New plants forming at the tips of stems; Strawberry, creeping buttercup.
Bulbs: Swollen underground buds; onion, bulbous buttercup.
Corms: Another type of swollen stem; onion couch (a grass weed).
Cuttings: Parts of stems which grow; blackcurrants.

PLANT FOODS

Animals must eat food which comes from living things; *organic* material from plants or from other animals.

Plants need foods which are quite different; *inorganic* substances such as gas (carbon dioxide) and soil minerals (phosphates and potash). Apart from water, the greater part of a plant is made from the air. The carbon dioxide which is important to plants is a waste product from animal life—breathed out from the lungs.

The green plant is a sort of machine operated by sunlight which grows and builds up food supplies from the air and the soil. Water is essential to a plant to make its own body and to move the foods inside it.

The Needs of the Plant

These are the chief food substances needed by the plant:

Foods	*Where From*
Oxygen, carbon	Carbon dioxide (CO_2), in the air
Oxygen, hydrogen	Water (H_2O)
Nitrogen, phosphorus, potassium, calcium, magnesium, iron, sulphur	In the soil
Nitrogen (with plants of the legume family)	Soil air
Boron, manganese, copper, zinc, cobalt, molybdenum	In the soil

How the Plant Makes Food

Water is passing through the plant all the time. It is taken in through the root hairs and goes up the stem to the leaves. On the surface of the leaves it is evaporated (transpired) into the air through the *stomata* which control the flow.

This stream of water which is moving all the time carries the food substances to the leaves. The amount of water any plant needs is very large; in a normal season in England it is often more than 200 times the amount of dry matter in the whole plant.

When there is a shortage of water, and the plant cannot get enough to replace what the leaves transpire, the plant *wilts*.

In the leaves, carbon dioxide which is taken in from the air through the stomata, is combined with water from the roots and with the plant foods brought up in this water, to make sugars, starches and other compounds. These products are used by the plant for ENERGY, GROWTH, STORAGE.

This production of food in the leaves needs the action of sunlight on the green colouring of the leaves (chlorophyll). This is known as *Photosynthesis*. Plants without chlorophyll (such as fungi) cannot make their own food and must feed on other plants or animals, or decaying material.

Very simply, this is how the plant makes food:

Water + Carbon Dioxide = Sugar (and oxygen given off as a by-product)

$$6H_2O + 6CO_2 = C_6H_{12}O_6 \ (+6O_2)$$

How the Plant Uses its Food

To find the energy needed for the work which it must do the plant breaks down some of its food: it 'burns up' sugar with oxygen taken in

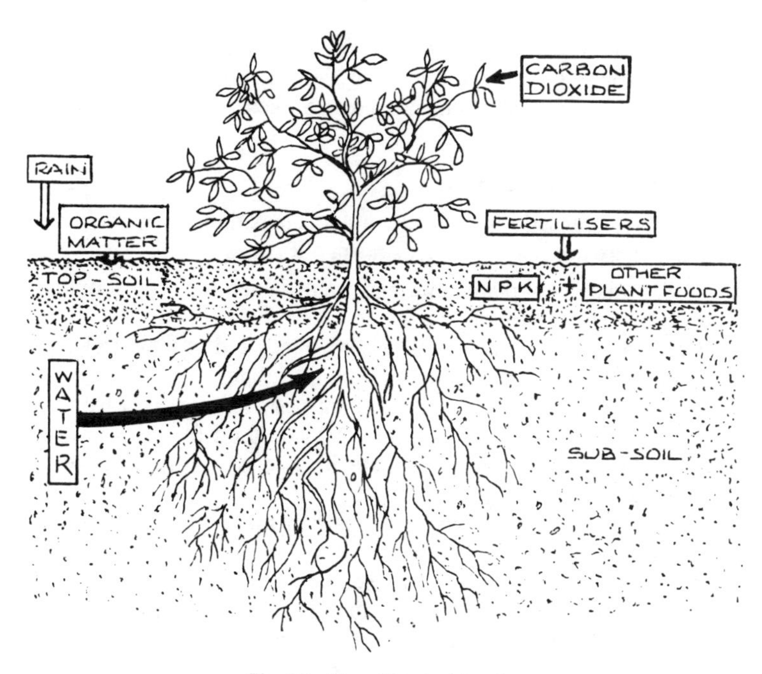

Fig. 10. How the plant feeds.

through the leaves. This is how the plant produces energy by the process of respiration:

Sugar + Oxygen = Energy (and water and carbon dioxide given off as a by-product)

$$C_6H_{12}O_6 + 6O_2 = \text{Energy} \ (+6H_2O + 6CO_2)$$

The plant uses its food supplies for these purposes:

1. To grow;
2. To store for future use (as in a seed or a storage root);
3. For maintenance—to keep all its systems working.

PLANTS IN THE WRONG PLACES—WEEDS

Weeds are so common that they are found on every farm and in every field. It is impossible to find land completely free from weeds. Many of these plants are well known, but you will find that they have local names which are often used much more than the 'official' names given in the reference books.

Any plant in the wrong place can be considered as a weed. Sometimes a crop plant is found growing at the wrong time in another crop and where it causes any trouble it must be considered as a weed.

Trouble Caused by Weeds

It is not always realised just how much damage is done by weeds and how many ways there are in which weeds harm our crops.

1. Weeds take away plant food and water which the crop should have, and thus reduce growth and yields.
2. Weeds may choke or strangle a crop (bindweed or cleavers), shade it from the light and may smother young seedlings.
3. Weeds may provide a home for plant pests and diseases from which they may infect a crop; they may provide food for pests or diseases during the winter; a crop choked with weeds is more likely to suffer from some fungus diseases.
4. Weeds in the straw of corn crops can make harvesting slow and difficult.
5. Weed seeds harvested with the seed of crops may spoil them for sale, make them unfit for feeding or cause expense in cleaning.
6. Some weeds are poisonous (ragwort, horsetail) and some cause taint in milk or the flesh of animals (wild onion, garlic, mayweed).

How Weeds Spread

Part of the danger of weeds in farming is that they can spread quickly and in some cases spread far and wide. These are the main ways in which weeds spread:

1. By creeping under or over the ground: thistles, couch grass, buttercups.
2. In the seed of crops: cleavers, wild oats, docks.
3. In crop remains which are spread or used for litter or feed for livestock: fat hen (weeds may spread in farmyard manure).
4. By wind: thistles, ragwort, poppy.
5. By birds or animals: wild oats and other grasses.

PLANT PESTS

There are many pests which will attack crops, but although little-known pests may be found causing damage at some time there are only a few agricultural pests of real importance. Insect pests have complicated 'life cycles' and with some of them it is important to understand their development so they can best be controlled. Some of the causes of pest damage to crops:

1. Bad growing conditions; any crop growing poorly or checked in its growth is more likely to be damaged by pests.
2. Susceptible crops; some crops are more likely to be damaged than others.
3. Conditions which help the pest to get started in land or a crop, or which encourage it.
4. Leaving hiding places, weeds or crop remains which help the pests to live over from one year to the next.
5. Bad rotations or no rotations.

Types of Insect Pest Damage

(a) *Biting*. These pests have biting mouth-parts and either eat the leaf, stem or roots of the crop so destroying it or weakening it; or damage it so that it fails to develop or is weakened, which may allow other pests or diseases to cause more damage. Examples: wheat bulb fly, wireworm, flea beetle.

(b) *Sucking*. These pests have sucking mouth-parts and weaken crops by settling on them and sucking their juices. Example: aphis on peas and beans.

(c) *Spread of Disease*. Some of the sucking pests spread virus diseases by sucking the sap from infected plants and carrying the virus to healthy crops. When they start to feed on the crop they put the virus into the plant sap. Example: aphis on potatoes and sugar beet.

How Insect Pests Live and Spread

Insects have two main types of life cycle, and it is useful to know the main stages:

(a) *4-stage*. This is the normal type of development for most of the best-known insects—flies, beetles and caterpillars. The *egg*, usually laid in very large numbers, may be laid on plants and on or in the soil; it may hatch very quickly or may stay for a time until conditions are right for hatching. The *larva* (often called a grub, maggot or caterpillar) hatches from the egg and usually grows fast and so needs plenty of food; this is often the stage when most damage is done. When the larva is fully fed it goes into a *pupa*, a sort of case in which it changes completely; it does not move about and is usually found in the soil. The *adult* hatches from the pupa for a short life when it mates and lays the eggs. Examples: wheat bulb fly, cabbage caterpillar.

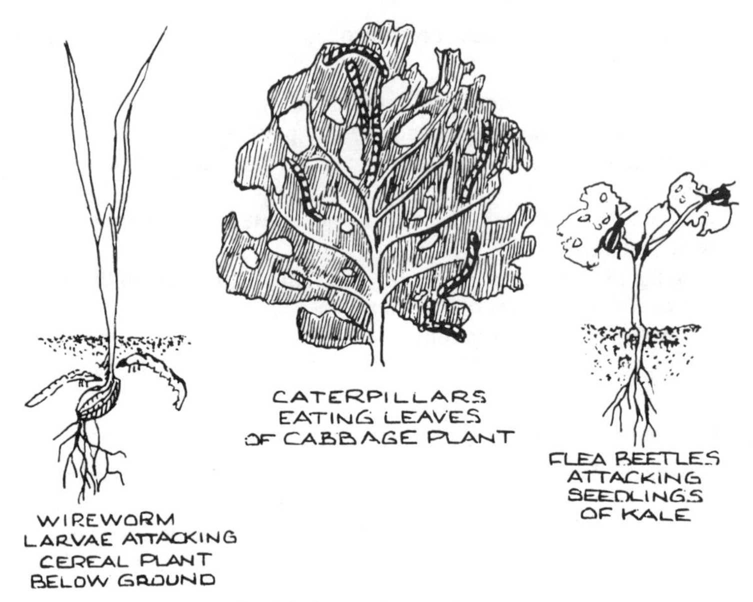

Fig. 11. Insect damage to crops.

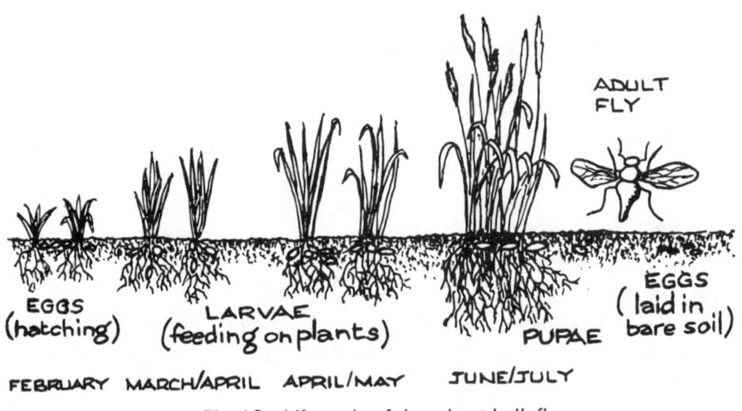

Fig. 12. Life cycle of the wheat bulb fly.

(b) *3-stage.* The eggs hatch out into *nymphs* which are rather like the adult but smaller. The nymphs feed and develop, sometimes getting rid of several skins as they grow. When fully fed the nymph has passed into the *adult* stage and the life story starts again. Examples: slug, aphis.

Aphids have a 3-stage life cycle but sometimes it becomes rather more complicated than this. Different types of generations develop during the season, some which fly and others which stay in the same place and some of them can produce young without mating.

Eelworms have their own peculiar form of life cycle. The eggs hatch in the soil and the young female larva gets to the root of a susceptible plant and feeds on it by sucking. The male fertilises her and she then produces eggs which she keeps in her body, swelling up until by the end of the season she becomes just a bag full of eggs, called a cyst. The cyst breaks in the soil releasing the eggs.

Important Plant Pests

Soil Pests. Some insects live in the soil during most of their lives and damage susceptible crops which are grown. In grassland (usually old permanent grass) click beetles lay eggs which hatch into wireworms and crane flies lay eggs which hatch into leatherjackets. Both these pests feed on the grass roots and, if the grass is ploughed, feed on most crops which may be grown.

Slugs live in the surface layers of the soil and are found mainly on heavy wet clay soils; they attack autumn-sown corn crops and in some winters cause a tremendous amount of damage.

Eelworms live in the soil but are different from the other soil pests because they only attack certain crops; they are known as specific pests. Potato eelworm only attacks plants of the potato family and beet eelworm only the beet and cabbage families. These pests increase if any susceptible crop is grown too often in any field and when present in large numbers are impossible to control and take many years to reduce in numbers.

Pest of Root and Green Crops. Flea beetle is commonly found in rough grass, ditches, etc and moves to crops of the cabbage and beet family as the seedlings start to grow. The beetles eat the seedlings before they come through the soil or just as they appear and continue feeding on the small plants. This is the period when most damage is done; eggs are laid and the larva feeds on roots during the summer. This pest lives over the winter as a beetle.

The cabbage white butterfly lays eggs on plants of the cabbage family. The egg hatches into a caterpillar which feeds and grows quickly on any crop of this family.

The Colorado beetle does not live in this country because it has been very carefully controlled. The larva has a tremendous appetite and will eat off the tops of potato plants. If this pest appears *it must be reported to the police at once.*

Aphids (green fly) live over the winter on crops of the cabbage family and move on to potatoes and beet crops in the spring. They cause damage to the crops by feeding on the plants, but their chief importance is in spreading the virus diseases.

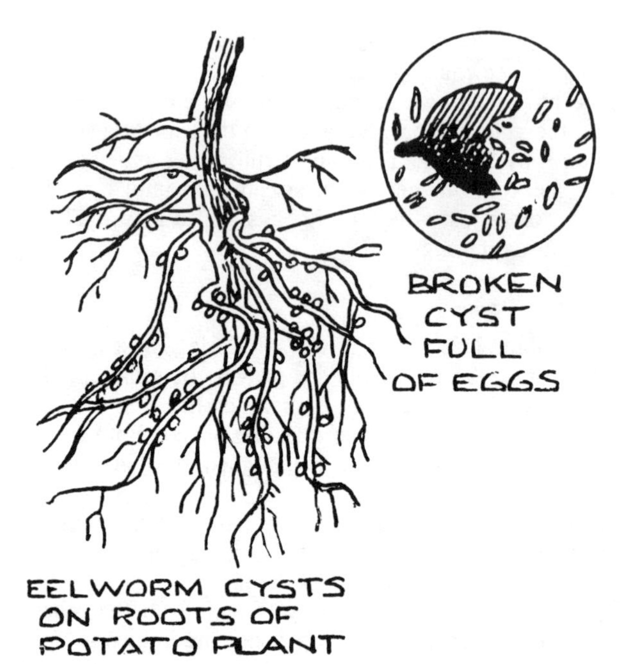

Fig. 13. Eelworm cysts on roots of potato plant.

Pest	*Crops Attacked*	*Stage of Insect Life when Most Damage is Done*	*Control*
Wireworm	All crops	Larva	Soil insecticides
Leatherjacket	All crops	Larva	Soil insecticides/baits
Slug	Mainly winter	Nymph/Adult	Baits + Draza pellets
Eelworm	Potatoes/beet/ cereals/peas	All stages	Rotation or soil insecticides
Wheat bulb fly	Wheat (and other cereals)	Larva	Rotation/seed dressing
Frit fly	Oats (and other cereals)	Larva	Early sowing
Gout fly	Barley (and other cereals)	Larva	Early sowing

Pests of Other Crops. Pea and bean weevil attacks by eating out notches around the leaf edges, weakening the crop. They lay eggs and the larva feeds on the root nodules of these crops. This pest winters as a weevil (adult).

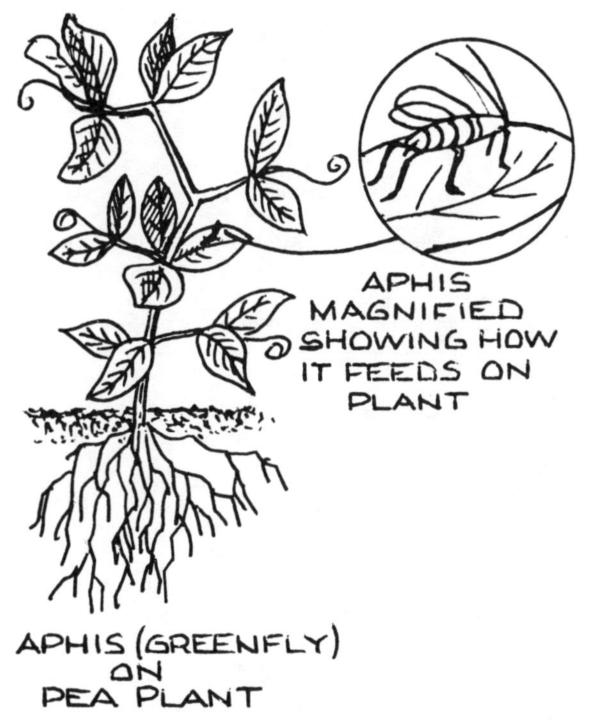

Fig. 14. Aphis (greenfly)on pea plant.

Pest	*Crops Attacked*	*Stage of Insect Life when Most Damage is Done*	*Control*
Flea beetle	Cabbage/beet families	Adult	Seed dressing/ insecticide on seedlings
Cabbage white butterfly	Cabbage family	Larva	Insecticide on leaves
Colorado beetle	Potato	Larva	Inform the authorities —(insecticide)
Aphids	Potato/beet/ cabbage family	All stages	Systemic insecticides

Pea moths lay eggs in the flower of the pea crop and the larva hatches and feeds as a small maggot in the green peas inside the pod, which spoils the crop for human use. The fully fed maggot falls to the soil, and the pupa lives in the soil over winter and hatches out as a moth in the spring.

Aphids are found on peas (green fly) and beans (black fly) and reduce the crop by feeding on it, often in large numbers.

Pests	*Crops Attacked*	*Stage of Insect Life when Most Damage is Done*	*Control*
Pea and bean weevil	Peas/beans	Adult (weevil)	Insecticide on leaves
Pea moth	Peas	Larva	Insecticides on flowers
Aphids	Peas/beans	All stages	Systemic insecticide

Other Pests. Insects are not the only pests attacking farm crops. Birds and mammals are also important and in some seasons and some districts cause a tremendous amount of damage.

PLANT DISEASES

There are a tremendous number of diseases and similar troubles of all plants, weeds as well as crops. Fortunately, only some of them are important at any one time or in any one district but as conditions change new troubles and diseases come along. It is important to understand the main troubles and to know how they can be avoided or controlled. Plant diseases may just reduce yields or spoil the quality of produce, or they may cause crop failure. Disease is more likely when large acreages of one crop are grown all together.

These are the main causes of plant disease:

1. Bad growing conditions—ill-drained land, shortage of plant foods, poor tilth. Poor crops are always liable to attack.
2. Susceptible crops—some crops and some varieties of plants are *resistant* to attack (will stand up to it) and some are *susceptible* (likely to be attacked).
3. Conditions which favour and help the disease spreading or getting established—hot wet weather, a late season which hinders crop growth, other checks to the growth of crops.
4. Wrong balance of plant foods; too much nitrogen.
5. Bad rotations or no rotations.

Agents of Plant Disease. If the conditions are there for a crop to become infected by disease, it will become infected if the *agent* of the disease is about at the time; this may be a fungus, bacteria or virus, all of which live as parasites, preying on other living things.

Fungus

A fungus is a living plant, usually very small. It has no green chlorophyll (see page 34) and so has to feed on other living things (plants or animals) or on decaying material. Threads of the fungus (called *mycelium*) get into various parts of the body of plants where they develop. At some stage in their life they produce *spores*, which are like the seeds of the fungus. These spores are produced in great numbers and spread the disease. If we know something about the spread of the disease, we may be able to control it.

(Note: we talk about a *fungus*; a number of *fungi*. A *fungicide* is a chemical which will kill fungi).

Soil-borne Fungi. Spores drop off infected plants, get into the soil and may stay there for a long time. When a susceptible crop is grown it is infected from the soil, producing more spores of the disease which are returned to the soil ready to infect other crops later. Examples: Take-all and 'Whiteheads' of corn crops.

Seed-borne Fungi. When the spores are free they become connected in some way to the seed of the plant they are attacking. This happens in two ways. When the seed germinates and grows, the fungus develops with it and may kill the seedling; or it may grow inside the plant causing the disease and in time producing more spores which infect the seed of the crop. The spores are carried within the seed in two different ways: *Either* on the seed (like covered smut of wheat) or inside the seed (like loose smut of wheat).

Wind-borne Fungi. When the spores are released, they are carried through the air on the wind until they settle on the leaves of a susceptible crop which is then infected. Some of these diseases are encouraged by wet and warm conditions which help them to develop. Example: Potato blight.

Insect-borne Fungi. The spores are carried on or in insects which take them from infected plants to other susceptible plants.

Bacteria

These are very small plants (see page 34) and the types causing disease are parasites on other plants. Bacteria reproduce by splitting and so can increase very quickly. Bacterial diseases are found in fruit and vegetable crops. Example: Fire blight of pears.

Virus

A virus is a living chemical, which cannot be seen under the microscope, and many viruses cause plant diseases. Viruses do not produce any form of spores but are themselves spread from one plant to another, often by insects. Example: Virus yellows of sugar beet.

Important Plant Diseases

Cereal Crops are attacked by fungus diseases which are spread through the soil, the seed and also by wind.

Take-all attacks wheat and barley, and is worse on light soils in a poor state of fertility; it causes the crop to die in patches and also causes small or empty ears to develop. *Eyespot* attacks wheat and barley and is worse on autumn-sown crops, weakening the plant near ground level, so the straw cannot stand and the crop is laid causing loss and harvesting troubles. Both these diseases are made worse by growing wheat and barley too often and can be kept under control by proper rotation; fungicides can be used against eyespot. *Covered smut of wheat* causes balls of fungus spores to replace some of the grain and these spores are spread on to the healthy seed grain when the crop is threshed or combined. *Loose smut of wheat* makes the plant produce black dust (spores) in some of the ears, which infects healthy grain and gets inside it. *Yellow rust* infects the leaves and in a bad year may make them wither and so reduce the crop.

Disease	*Crops Attacked*	*How Spread*	*Control*
Take-all	Wheat/barley	Fungus in soil	Rotation
Eyespot	Wheat/barley	Fungus in soil	Rotation, fungicide
Covered smut	Wheat/barley	Fungus *on* seed	Seed dressing
Loose smut	Wheat/barley	Fungus *in* seed	Seed from healthy crops/chemical treatment
Rust	Wheat	Fungus in air	Resistant varieties, fungicide sprays

Root Crops are attacked by fungus, bacterial and virus diseases, and a great deal of care and money is spent to control these troubles. *Potato blight* is found in most districts, although it is much worse in wet areas and worst under hot damp summer conditions. *Leaf roll* and *mosaic* are virus diseases of potatoes which cause poor yields and crop failure unless healthy seed is got from certain districts such as Scotland and Ireland. *Virus yellows* affects sugar beet and other crops of the beet family during the summer, causing growth to stop early and thus poor yields. *Common scab* causes marking of the skin of potatoes, is worse on light gravelly soils and always worse directly after lime is added to the soil. *Dry rot* is a common disease of seed potatoes; some varieties suffer more than others and damaged potatoes are always likely to become infected.

Other Crops. Chocolate spot is a very common disease of field beans which damages part or all of the leaf and reduces yield *Clubroot* affects

Disease	*Crops Attacked*	*How Spread*	*Control*
Potato blight	Potato/tomato	Fungus in air	Fungicide spray
Leaf/Mosaic roll	Potato	Virus by aphis	Healthy seed
Virus yellows	Beet family	Virus by aphis	Systemic insecticide
Common scab	Potato	Fungus in soil	Organic matter in soil
Dry rot	Potato	Fungus on boxes gets into damaged potatoes	Disinfect boxes; treat seed; avoid damage

most crops of the cabbage family, and by causing ugly swellings on the roots weakens the plant and may kill it; this disease is worse on land which is short of lime. *Clover rot* attacks red clover, alsike clover and field beans; it makes the crop die out in patches.

Disease	*Crops Attacked*	*How Spread*	*Control*
Chocolate spot	Field beans	Fungus in air	Enough plant foods, especially Potash. Fungicides.
Clubroot	Cabbage family	Fungus in soil	Liming/rotation Chemical dressing of seedlings
Clover rot	Red clover	Fungus in soil	Rotation

Deficiency troubles often look like diseases and may affect the crop in just the same way as a disease. They are caused by a shortage of one or more of the plant foods and can usually be controlled by supplying that plant food (see page 34).

THINGS TO DO

1. Study seeds of different types, see how they germinate, and look at the various stages of growth. Collect small samples of seeds.
2. See how plants develop and spread by methods other than by seed; look at plants of couch grass, white clover, strawberries, onions, and potatoes.
3. Study and be able to recognise the common weeds of your district, and find out how they spread, and in which crops they are commonly found.
4. See examples of damage by common plant pests in your district, and if possible see the pests themselves at different stages.
5. See examples of damage caused to crops by plant diseases.

QUESTIONS

1. Which are the main parts of a plant, and what is the function of each part?
2. Which are the main plant families which are used by us for normal farming in this country, and which crops come in which plant family?
3. What are the main methods by which plants reproduce and spread?
4. What are the main insect pests which affect farm crops in your area at the present time, and what damage do they do?
5. What are the main diseases of farm crops in your area at the present time, and what harm do they cause the crop?

Chapter 3

Farming

The Scientific Use of Soil and Plants

MANY IMPORTANT things in farming can be controlled by the farmer; these include the condition and the fertility of the soil, within certain limits, and the growth, health and quality of crops. To do this properly the farmer uses his experience and technical knowledge and calls on the help of the specialist and the scientist.

Later in this book, something about the science of the soil and the workings of plants is explained in simple terms. Such knowledge is useful, but it is essential all the time to relate it to the realities of farming practice. Farming is a collection of skills, a science—some would say an art also—but above all today it is a business.

KNOWING THE LAND

No farmer in his right mind would try to grow malting barley on the slopes of Snowdon, nor would he make a living running one poor ewe per acre on rich fen land. The farming must be suited to the natural conditions. It is important to know something of the land and types of farming which are best done in different parts of the country; the land and types of farming in your own county (and in your own parish); and the reasons why such differences are found.

Farming well means making the best (and the most scientific) use of natural conditions, land, crops, livestock, machinery and all the other things which have to be put together to make the farming system work. Without a full and realistic knowledge of most of these, it is difficult to make the right decisions.

Important points to know about your district, and about the country as a whole, are these:

Height of Land (Altitude)

The higher the land, the poorer it becomes and the heavier is the rainfall; conditions are harder and the land is exposed to extreme conditions. This makes possible only a limited type of farming, not very productive, and restricted in the livestock which can be kept and the plants which can be grown.

Rainfall

The main winds found in the British Isles (the prevailing winds) come from the West, carrying damp air from the Atlantic Ocean. Much of the rain formed from the moisture in the air is dropped in the mountainous districts of the West, and weather conditions become drier as we move over to the East.

The wet western districts grow much more grass—in some parts the land may be nearly all grass—and the Eastern Counties are used mainly for arable farming. Rainfall is measured by the amount of rain falling on the land, and recorded in a rain gauge. The rainfall per year varies from 2000 mm (80 in) in the Welsh mountains to 500 mm (20 in) on the Essex coast. In the dry East, there is not enough rain to grow the largest possible crops, and irrigation is sometimes used to make up for this water shortage (or deficit).

Temperature

The higher and wetter a district, the colder it is likely to be. Near the sea, the climate and temperature of the land are usually affected and are less likely to be extremely hot or cold. Parts of the country are affected by warm currents in the sea (the Gulf Stream) and this encourages early crops.

Aspect

The way the land faces, the compass direction of its slope, affects it for farming. Land facing North is later to warm up and start growing. Land facing South is always kinder and earlier.

Land on slopes from which the air drains away slowly or not at all can cause trouble. This may form a 'frost pocket' in which frost collects and will damage crops such as potatoes or vegetables. The problem is worst for fruit growers but sometimes has to be considered on a farm.

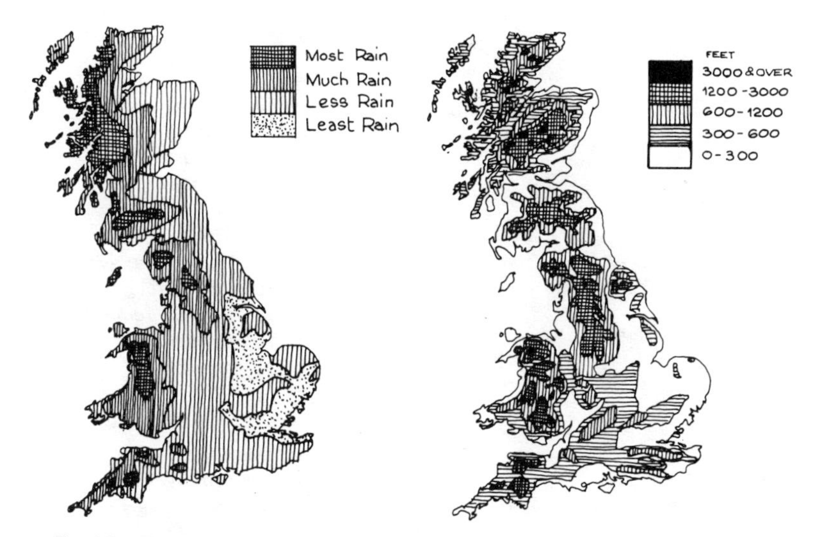

Fig. 15a. Pattern of rainfall in the British Isles.

Fig. 15b. Height of land in the British Isles.

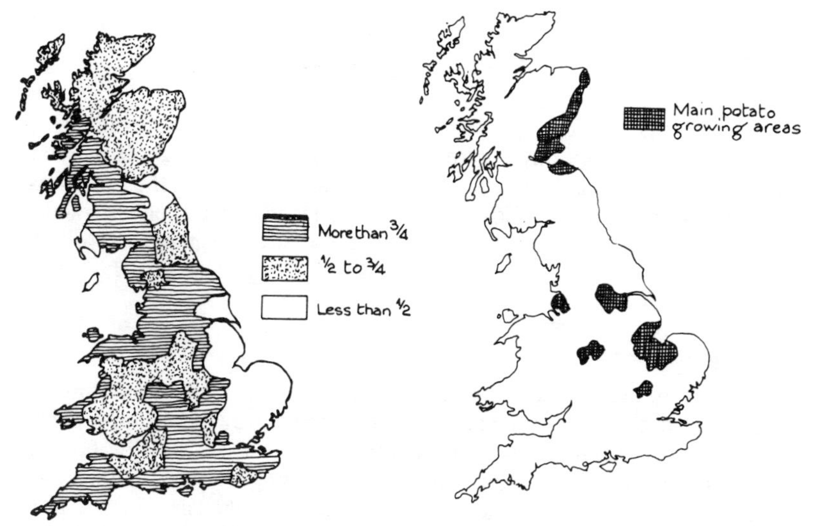

Fig. 15c. Areas of grassland in the British Isles.

Fig. 15d. Main potato-growing areas in the British Isles.

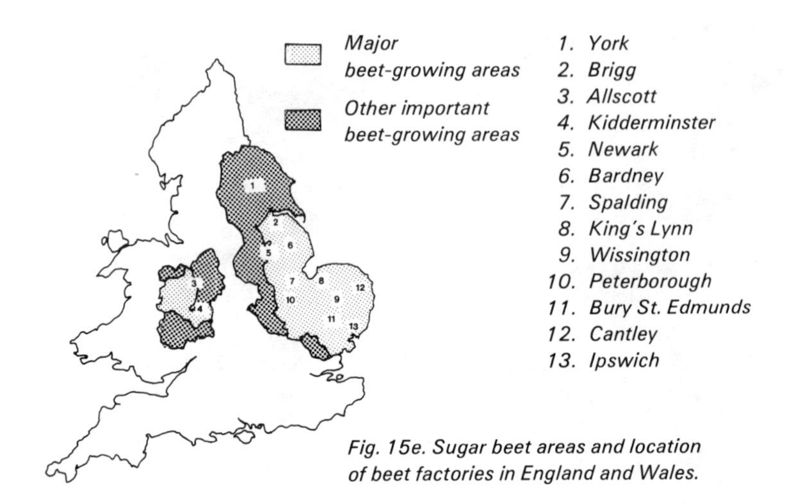

Fig. 15e. Sugar beet areas and location of beet factories in England and Wales.

Soil Types

This is not an easy matter and needs to be studied and understood in some detail (see page 15), but there are some well-known areas of the country which have distinct and well-known soil and farming types. This may be due to the rock material which lies under the soil and which influences everything above it.

KNOWING YOUR FARM

Any farmer thinking about taking over a new farm, examines it thoroughly before he buys or rents it—his living depends on it. Anyone starting to work on a farm needs to get to know the farm in the same way. What the farmer sees and how he looks at the land depends on his training and experience. There are things to look for, tests to make, records to keep.

Walking the Land

There is an old saying: 'The farmer's foot is the best manure.' It means that detailed care and management, the farmer prowling around seeing and knowing what is going on, means more to the success of the farming business than any other single thing. Knowing your land in this way, walking it, and going right into your crops, leads to the care and the good practical management that can properly be called *cropmanship* (similar to the stockmanship that is needed with animals). It gives you a chance to see how the land lies, the different types of soil, the patches of poor land, the weeds that mean trouble or previous bad management, the blocked drains and the first signs of pests and diseases.

Handling the Soil

When dealing with soil, after the foot the hand is the next best testing tool. There is a whole method of soil testing, putting your soils into their proper classes, based on handling the soil (see page 15). Learn to know the difference between the main types of soil, sandy, silty, loam or clay, and also the difference between a soil in good order and one which has been mistreated or neglected. Learn to handle soil so that you can know when it is dry enough and when it is moist enough for certain jobs to be done.

Soil Sampling

For some purposes, soil samples must be taken and in many cases sent away for analysis in a laboratory. Sampling is usually done with an auger or with a trowel; the samples must be done to the right depth (usually the topsoil is being sampled) and the 'cores' of soil must be mixed together thoroughly.

Drainage

Where lands needs drainage this can often be seen clearly, but sometimes the signs of waterlogging are hidden and must be looked for (see page 54). Land that is too wet cannot do its best; crops are poor and other troubles can follow.

Maps

Good maps are essential. Those commonly used for farming purposes are the ordnance maps, produced for sale by Her Majesty's Stationery Office. Farm maps and plans are based on these ordnance maps. The farmer uses these maps, and extracts from them, to record cropping and drainage plans. You should get to know the different types: 1:25,000, 1:10,000, and 1:2,500 which is used to provide detailed maps of farms and fields.

Cropping Records

Good records are vital. No man's memory is clear enough to give a true picture of details, even if it was only last year. There is so much detailed technical information in farming that proper records must be kept. For this purpose, a good farm diary is the first essential.

Observation

Farming is a job that makes you keep your eyes open if you are to do it well. The agricultural industry needs people who can do this, who can observe. There is always something of interest happening on a farm, whether in the fields, around the buildings or among the livestock. Note everything that happens, think about it, talk about it. In this way you increase your own knowledge and you make yourself a much more useful person on the farm.

SOIL FERTILITY

Soil fertility is a measure of the ability of the land to produce crops. Some soils are naturally highly fertile; the Fens, and some good clay and loamy soils. Some soils are naturally poor and hungry; heath soils and other light sands.

Fertility is a combination of several different things in the soil. All are important and if any one is low it will prevent all the rest being used to the full, just as in the old saying that a chain is only as strong as its weakest link.

The various things that make up soil fertility are:

1. Moisture conditions—water in the soil.
2. Chemical conditions—plant foods and lime (N, P, K, etc).
3. Soil structure—organic matter and cultivations.
4. Soil life (bacteria etc).

All these things are more or less under the control of the farmer. In the following pages their control is considered.

CONTROL OF SOIL CONDITION

One important job for the farmer is to get his land into good condition, and to keep it that way. Land in good order works better, can be cropped with less trouble, provides better conditions for germination and crop growth; in the end it produces better crops. The condition of the land is controlled in two main ways.

Cultivation

The test of a good arable farmer is his cultivation, which depends on a thorough knowledge of his land and his ability to cultivate it in the proper way and at the right time. Timeliness, which means doing any job at the right time, is one of the most important things in handling soil cultivation. It is much more difficult on the heaviest clay soils to do everything just right; much easier on a light sandy soil which can be worked at almost any time.

We cultivate the land to:

1. Provide a proper seedbed for crops and proper conditions for them to grow and develop;
2. Control soil moisture and soil air;
3. Destroy weeds;
4. Bury rubbish (trash) and to mix in manures and fertilisers;
5. Control pests and diseases.

Cultivating stirs up the soil, breaks the furrow slices, knocks lumps and clods about and breaks them down, and loosens the soil when it has become solid. A cereal stubble is sometimes cultivated (this is known as stubble cleaning) to make weed seeds grow, before the land is ploughed.

Harrowing breaks the land down to a fine condition after cultivating, covers seeds, breaks a soil cap and destroys weeds.

Rolling is used to make the land firm, to push stones down into the surface and break clods.

This list shows all the jobs which may need to be done after one crop is harvested until the time the seed is sown for the next crop:

1. Stubble cleaning or cultivating.
2. Ploughing (sometimes land is ploughed a second time).
3. Liming.
4. Heavy cultivation or discing.
5. Spreading fertilisers.
6. Harrowing.
7. Rolling.
8. Sowing the seed.
9. Seed harrowing.

Organic Matter in the Soil

Organic matter helps to keep the soil in good condition. It holds moisture, gives body to light land and opens up heavy land so that it is easier to work. Land which has too little organic matter loses condition, becoming lifeless and difficult to work. On some soils which have been cropped hard for years there is sometimes a danger that there is too little organic matter.

In farming practice, soil organic matter should not get below a certain level in any soil to avoid running into problems. At the present time this should be about 3 per cent, but it must vary according to the type of soil and the way it is farmed.

Organic matter is added to the soil in normal farming in these ways:

1. Ley farming—putting the land down to grass for three or more years and then ploughing in the grass for a few years' arable cropping.
2. Crop remains—roots of all crops, straw, tops of root crops.
3. Farmyard manure—which is a mixture of straw, dung and litter.
4. Green manuring—growing green crops such as mustard for ploughing in. This is not often done today.

CONTROL OF SOIL WATER

Too much water in the land causes far more trouble than not enough water. Even in the dry Eastern Counties, land drainage is very important and there is still much land which is not doing its best because it is badly drained. Shortage of water for crop growth is also a problem, but it does not concern so many farmers.

The object of land drainage is to remove surplus water, usually by controlling the level of the water table (see page 55). If there is too much water in the soil, there is not enough soil air and this is bad both for the plants and for other living things.

These are the troubles caused by bad drainage:

1. Poor root development—and a poor plant.
2. Wet land is late land—slow to grow crops in the spring.
3. Wet land is cold land—this holds back the growth of plants.
4. Bad soil structure—the land is difficult to work.
5. Manures and fertilisers are wasted or not used properly.
6. On wet land, soil troubles and diseases of crops and livestock are much more common.

It is not always obvious that land is suffering from bad drainage but these are some signs that land is badly drained:

1. The land lies wet for long periods.
2. The land dries out in patches during the summer.
3. Poor patches in crops, often stunted growth, and yellow-looking plants.
4. Coloured markings in the soil, usually streaks of red and brown, due to iron salts and lack of air.
5. Particular weeds, such as horsetail, coltsfoot, rushes, sedges, and watergrass.
6. Water coming to the surface; wet and spongy soil.

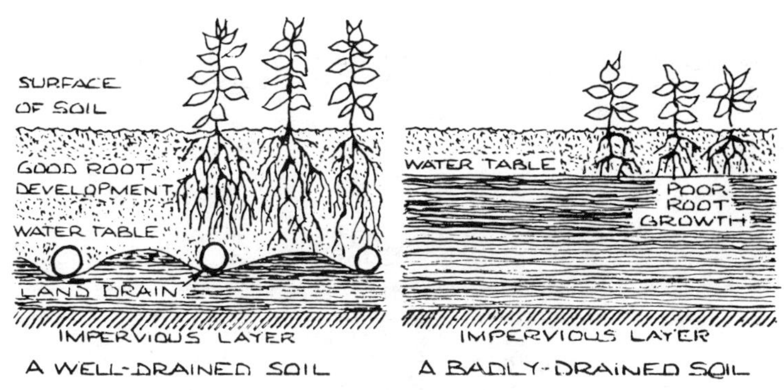

Fig. 16. Land drainage affects the growth of crops.

Draining the Land

Land drainage is a complicated and expensive job which must be planned and carried out properly. It is usually done by contractors who specialise in this work. Government grants are paid for land drainage and drainage officers of the Ministry of Agriculture advise on drainage problems and inspect the work.

There are several methods of draining land which aim to get the water away to the main watercourses.

Ditches. It is important to know who a ditch belongs to. Ditches must be kept clear of rubbish, running freely with the banks and bottom in good condition and below the level of the drain outfalls.

Tile Drains. Tiles can be used in any type of soil. The layout is planned first, then trenches are cut to allow a proper fall (so that the water runs properly), tiles laid in the bottom, covered with a layer of clinker or shingle and the soil filled back. Tiles are made of porous earthenware, concrete or plastic, and in different sizes, 10 or 15 cm (4 or 6 in) in diameter.

Tiles are put in at a depth of 1 metre (39 in) or more so that mole drains can be put above them and will not be disturbed by any cultivation. These figures give some idea of the size of outfall and main necessary to drain a field:

10 cm (4 in) main will take outflow from 2 hectares (5 acres)

15 cm (6 in) main will take outflow from 6 hectares (15 acres)

Long plastic pipes with slits in them are now being used, and it is possible now to drain through the growing crop.

Mole Drains. Moles can only be made in heavy land, with a clay subsoil free from stones or veins of sand. A mole plough is pulled through the soil making a tube-like space in the clay which may be 5 to 8 cm in diameter. These moles are made at 45 to 60 cm deep and usually at about 3 m apart.

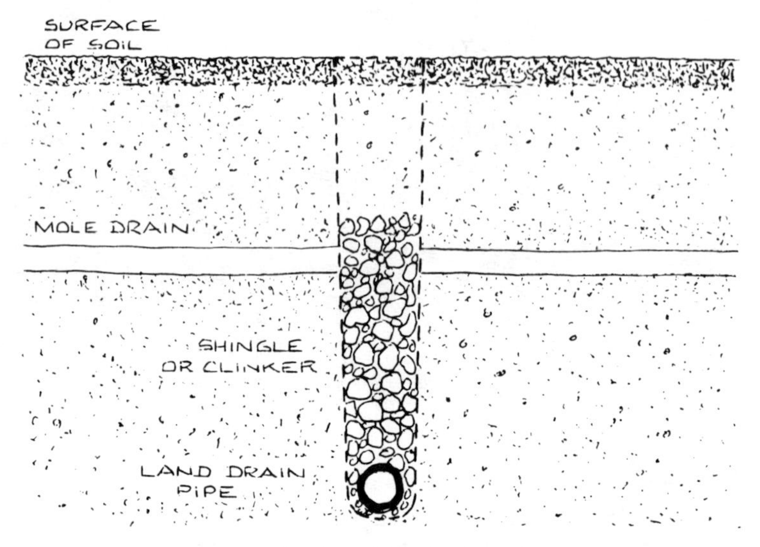

Fig. 17. Land drains in clay soil.

It is common to make moles up and down the main slopes in a field and running over the tile drains so that the water gets away properly.

As well as leaving channels in the soil, one effect of moling is to split or fissure the soil from the mole up to the surface which helps water to get away more quickly. Moles will last up to seven years normally, even longer in very heavy land.

Irrigation

In the drier eastern and southern counties of England, there is usually a shortage of water for crop growth during the growing season of the year. Obviously it increases crop yields to apply water, but this is expensive. The heavier soils hold water and thus can keep crops going in a dry time so irrigation is used mostly on the lighter land. As it is expensive, it is carried out mainly on larger farms and on crops which will give a high cash return: potatoes, sugar beet and vegetable crops.

CONTROL OF SOIL ACIDITY

Soil Testing

The acidity (sourness) of the soil can be measured by testing. This is shown by the pH of the soil. High pH (7) means plenty of lime, and low pH (5 or less) shows that a soil is very acid. In practice soil acidity can be measured in the field by the farmer or a technical adviser. This is done on the spot, using simple apparatus which shows a colour change; red for acid, green for plenty of lime. A more accurate method is to take soil samples away and test them in the laboratory.

Signs of an Acid Soil

There are some signs that a field is short of lime, either in patches or all over, which include:

1. A mat of dead stuff in grassland; no earthworms.
2. Poor growth of clovers.
3. Failure of certain crops, particularly barley and sugar beet.
4. Club root disease in brassica (cabbage) crops.
5. Certain weeds, sorrel (sourdock), spurrey (sand-weed), gorse, foxgloves, bracken.

Effect on Crops

Crops vary in the amount of acidity they can stand in the soil and most crops prefer less acid conditions. The crops more sensitive to acid conditions are:

Barley, red clover, peas, mangels, sugar beet and lucerne, all of which will fail or grow very poorly if the soil is acid.

pH	*Description*	*Crops which will Grow*
Below 4·8	Extremely acid	No normal crops grown in the British Isles.
4·8 to 5·2	Strongly acid	Oats, potatoes, kale, rye, ryegrass, field lupins.
5·3 to 5·7	Moderately acid	Those listed above plus wheat, beans, swedes, turnips, white clover, maize, cabbage.
5·8 to 6·4	Slightly acid	All normal crops
6·5 to 6·9	Very slightly acid	All normal crops
Above 6·9	No acidity	All normal crops

Lime Requirement

This is the amount of ground limestone or ground chalk needed to correct the acidity of the top 15 cm (6 in) of a soil to make it suitable for general cropping and to bring the pH figure up to 6·0–6·5. When soil is tested, this figure is given in terms of tonnes of ground limestone (calcium carbonate) per acre (or per hectare). The amount needed varies according to the type of soil; sour heavy land needs more lime than sour sandy soil.

Remember also that calcium is a plant food, needed in varying amounts by most crops.

Types of Lime

Several different types of lime are available, and in your district you may find there is a choice. In this case, it is wise to compare prices and values. Other forms of lime can be used so long as you adjust the quantity to that needed. These figures show how other forms of lime compare with 1 tonne of ground limestone:

Hydrated (slaked) lime	Calcium hydroxide	0·75 tonne
Ground chalk	Calcium carbonate	1·0 tonne
Lump chalk or limestone	Calcium carbonate	1·1 to 1·5 tonne
Waste (factory) lime	Calcium carbonate	1·5 to 2·0 tonne

Fig. 18. Loss of lime from the soil.

Hydrated lime is too expensive for farm use; it is sometimes used for the garden. Ground chalk and ground limestone are commonly used and can be put on the land at any time without harm; lump chalk or limestone is cheaper, but needs a hard winter's weathering to break it down. Waste lime (such as lime from a sugar-beet factory) is often cheap—sometimes free for the collection—but as it is wet and bulky it is too expensive to transport any distance.

How to Apply Lime

Large dressings of lime—more than 15 tonnes per hectare (6 tonnes per acre)—should not be applied all at once; better to give in two applications. The common forms of lime used on farms—ground limestone and chalk—can be spread at any time of the year. On grassland it is best to let the rain wash it down before livestock are grazed.

Any form of lime should be spread as evenly as possible over the land. It is best to lime land after ploughing, so the lime is worked into the top layers of the soil. Do not lime land before a crop of potatoes is grown, as it can cause scab disease.

CONTROL OF PLANT FOODS

The supply of the main plant foods varies with the type of soil (see page 15) and its previous cropping and management, but it is under the control of the farmer. With modern knowledge and with the fertilisers which are available any combination of the main plant foods can be added to the soil. As heavier yields are grown, due to modern varieties and growing methods, crops take from the soil larger amounts of the less important plant foods, some of which are only needed in small traces. Some of these shortages (deficiencies) cause trouble, but several of them are now under control and can be put right easily.

For growing conditions to be right for crops, there must be:

1. Enough lime.
2. Enough of the main plant foods.
3. Enough of the less important minor plant foods, or trace elements.

Main Plant Foods

The big three are nitrogen, phosphorus and potash—known as N, P and K.

Nitrogen is needed by the plant to increase growth, to increase yields and to increase the size of the leaves.

It makes the leaves of any crop a deeper green colour, and it can be seen easily where it has been used. Nitrogen produces quick results.

Too little N causes poor stunted weak plants, with yellow-looking leaves.

Too much N causes too much rank growth, soft plants which cannot stand properly, lodging (weak straw) in cereal crops and often makes plants more liable to disease.

N is needed particularly by any plants which make a good deal of leavy growth, grasses and forage crops, and for all heavy yielding crops.

Phosphorus is needed by the plant to develop root growth, to establish the plant when it is young and to help plants to ripen early.

P is important in the early stages of all plants, and is particularly needed for clovers and other legumes and for roots. It is easily locked up in some soils (this means that plants cannot use it) and for this reason it is better given 'little and often' rather than in large doses.

Too little P causes poor growth, and lack of clovers in grassland.

Too much P may cause crops to ripen too early.

Potassium is needed by the plant to keep it healthy, to improve quality and to resist drought.

K is needed particularly by plants which store sugars and starches, such as root crops.

Too little K causes poor growth, scorched leaf edges and diseases.

Too much K makes plans slow to ripen and may hinder germination.

Remember: N for yield and leaf growth
P for root development and earliness
K for health and quality

Signs of shortage: N—small crops, yellow leaves.
P—dull leaves, with purple patches, poor root growth.
K—scorched leaf edges, poor growth, diseases.

Less Important Plant Foods

Besides the big three (N, P, K), some other plant foods are needed in small amounts. On some soils and in certain districts, shortages (deficiencies) of these minor plant foods may cause trouble.

Magnesium shortage may be a problem on light soils in a wet district, especially if there is intensive cropping with cereals, and if organic matter is low in the soil. It also causes trouble to livestock grazing grass grown on the red sandstone soils.

Sodium is needed particularly by plants of the beet family, sugar beet and mangels. To get the largest possible yields it should be applied in the form of agricultural salt (sodium chloride) at 370 to 600 kilos per hectare (150 to 250 kilos per acre).

Sulphur is needed by most plants, but in practice it has been unusual to find any shortage in this country. Now, with the cutting down of industrial pollution of the air, a shortage of sulphur for crops is becoming more common.

Trace elements are needed by plants in very small amounts—it may be as little as a few grams per acre or even less.

Boron. In some districts, and if too much lime is used, a shortage of boron causes heart-rot in root crops of the beet family and in turnips and swedes.

Copper. In Eastern England, corn crops may be troubled by a shortage of copper on some light peaty soils, such as heaths.

Manganese. On Fen soils that are well-supplied with lime and with water. Oats sometimes suffer from grey-speck and peas from marsh-spot because of a shortage of manganese, and potatoes may also be affected.

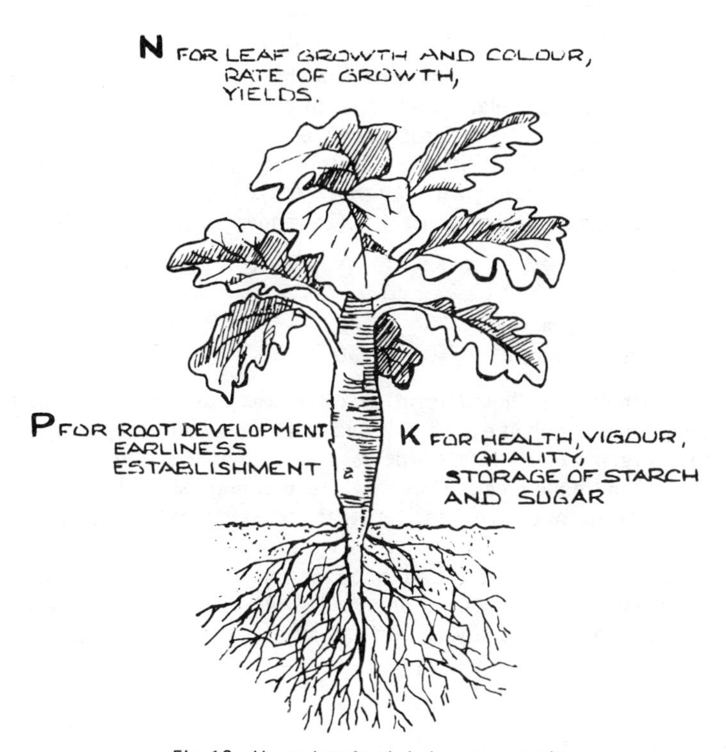

Fig. 19. How plant foods help crop growth.

Other minor plant foods needed in very small amounts are chlorine, zinc, and molybdenum.

To Avoid Trouble

Where a shortage of any one of these minor plant foods is expected, something can usually be done in good time. Magnesian is sometimes a convenient way of adding magnesium to soils which need it. Magnesium, manganese and copper can be given to the crop in spray form. Boron and magnesium are supplied in special fertiliser mixtures which can be ordered if required.

Supplying Plant Foods to the Crop

The only crops which supply part of their own plant food needs are the legumes: clovers, peas and beans. Through the nodules in their roots they take the nitrogen they need from the soil air; with it they feed not only themselves, but other plants growing with them (as in a mixed crop); this N can also benefit other crops grown in the following year (like cereals after beans). All other plant food needs of crops must be met either from the soil, or from manures or fertilisers added to the soil.

In the past, before 'artificial fertilisers' were made, the chief supply of plant foods was in farmyard manure (FYM). This is still used, and an average dressing of 10 tonnes per acre (25 tonnes per hectare) will supply:

	N	P	K	Magnesium
units per acre	30	40	75	15
kg per hectare	40	50	95	20

Slurry (manure in liquid form) is often used, and so also is poultry manure, which is rich in plant foods and can scorch some crops.

Other organic manures are little used now; industrial and food wastes, fish, meat, and bone meal, brewery waste, etc, may be used for vegetable crops if they are not too expensive. Near the coast, seaweed is sometimes used.

The main source of plant food is still the soil itself, and for this reason it is important that the soil is analysed from time to time. With the information from soil analysis, you can adjust fertiliser dressings applied to crops to get the best advantage.

Supply of Fertilisers

The fertiliser industry is well organised in this country, supplying a wide range of products designed to meet most crop needs. It also carries out investigation and research work, and supplies technical information. You can get lists of fertiliser recommendations from all the main manufacturers.

There are two main types of fertiliser:

(a) Compounds—very commonly used—which supply a mixture of two or more plant foods.
(b) Straights—which supply only one plant food.

Most fertilisers used now are compounds, many of them in granular form (in small fragments, easy running and easy to handle) and these supply the basic needs of most crops. The straights are mainly used in two ways:

(a) to top up the basic needs of a crop—for example, top-dressing winter cereal crops with nitrogen in the spring
(b) to put right major deficiencies in the soil—for example, a shortage of P or K in a particular field, or on a certain type of soil.

Using Straight Fertilisers

There are several common straight fertilisers, used for these and other special purposes. They may be mixed on the farm, but very little of this is done now (for a combination of plant foods, it is normal to use compounds). Straights available at the time of writing are shown below, but they may vary in form and in concentration from time to time. You should check with local suppliers to find out what is available.

	% of Plant Food	*Notes on Use*
Nitrogen (N):		
Nitram	34·5	
Nitro chalk	26·0	
Nitrashell	34·5	
Nitro-top	33·5	
Nitrate of soda	15·5	Very quick acting
Phosphate (P):		
Triple superphosphate	48·0	
Single superphosphate	20·0	
Slag	10–16	Special use on grassland
Potassium (K):		
Muriate of potash	60·0	
Sulphate of potash	50·0	Used for horticultural crops: expensive

Choice of Compound Fertilisers

The easiest method is to choose the compounds you need for your various crops from the range of fertilisers offered by one or more manufacturers. Thus, several different firms will supply suitable fertilisers for winter wheat, spring barley, oil seed rape or potatoes. The recommendations are normally in bags per acre, or bags per hectare; these are 50 kilo bags (near enough one hundredweight). The recommendations from the manufacturers give different rates, and different compounds, to suit the soil type and the level of fertility of the land.

Comparing Price and Value

With compounds, you can easily compare the price of similar compounds from different firms.

With straights, there may be different products to compare; for example, one firm offers a nitrogen fertiliser with 25 per cent N; another offers one with 33·5 per cent N. To compare the value, divide the price per tonne in each case by the percentage—this gives a 'unit price'.

Crop Needs for Plant Foods

To check accurately what fertilisers should be applied to any crop, plenty of information is available—some of it in much detail. There is advice from the fertiliser firms, the Ministry of Agriculture, the British Sugar Corporation; there are technical books, farming magazines, and the knowledge and experience of farmers. The figures given here (page 65) are *average* figures, shown in two forms which are both in use; they must be related to soil type, the season, previous cropping, and any deficiencies that may be shown by soil analysis.

Measuring Plant Foods

In recent times, farmers and advisers have come to use *units* of plant food per acre (a unit being about 0·5 kilo or just over 1 lb of any plant food). There is now a move to use kilos of plant food per hectare (kg/ha). Both systems are in use, and in this book both are shown.

Plant Food Ratios

Some of the figures shown in the table (page 65) can be converted easily to a simple base ratio. Thus 100:50:50 is 2:1:1 and 40:80:80 is 1:2:2. You will find that several of the compound fertilisers are made to fit these ratios, more or less. Sometimes these figures are marked on the bags.

Obtaining Fertilisers

After working out quantities of fertilisers needed, these are ordered and delivered. It may be necessary to store the fertiliser for a time. This must be done with care and forethought—so that the material can be used quickly and without too much moving about, and so that it is stored properly and safely. Store bags flat, clean and dry, away from water and preferably in a building. There is a fire risk with some fertilisers.

At the present time, most fertilisers are supplied in plastic bags. It is also possible to buy fertilisers in bulk and handle them with bulk equipment on the farm. A percentage of fertilisers are also supplied in liquid form; these are stored in tanks on the farm and applied to the crop with a sprayer. In the future, things may change—more of one sort, less of another.

Specimen Applications of Plant Foods in Fertilisers to Crops

	Units per acre			*Kg/ha*			*Plant food ratios*		
	N	P	K	N	P	K	N	P	K
Winter wheat	15	40	40	20	50	50	1	3	3
Spring barley: Feeding	80	40	40	100	50	50	2	1	1
Malting	60	40	40	80	50	50	1½	1	1
Oats	60	40	40	80	50	50	1½	1	1
Beans	0	40	40	0	50	50	0	1	1
Peas	0	40	40	0	50	50	0	1	1
Winter oilseed rape	50	50	50	60	60	60	1	1	1
Maincrop potatoes	176	200	240	220	250	300	1	1	1¼
Sugar beet	80	60	80[1]	100	75	100[1]	1	¾	1
Carrots	50	100	75	60	125	95	1	2	1½
Turnips and swedes	40	80	80	50	125	125	1	2	2
Kales	100	100	50	125	125	60	2	2	1
New ley	50	75	50	60	90	60	1	1½	1
Hay crop	60	30	30	80	40	40	2	1	1

Note: With winter cereals, a small amount of N only is given in the seedbed. Most of the N is applied in the spring as top dressing: 40–140 units depending on previous cropping. Similarly with winter oilseed rape: 160 units per acre.

[1] This assumes approx 3 cwt/acre of agricultural salt. Where no salt is applied increase potash by another 80 units/acre (125 kg/ha).

Applying Fertilisers to the Crop

Fertiliser must be put into the seedbed, where it will do most good to the young plant as it grows. This is done in two ways:

(a) *Broadcasting.* Fertiliser is spread with one of the several types of distributor (or sprayed in the case of liquids) on the seedbed before the seed is sown, and usually harrowed in.
(b) *Placement.* Fertiliser is put in at the same time as the seed and near to it in the soil, sometimes with it, and often alongside or even below it. For cereals, a combine drill is commonly used.
(c) *Top dressing* means putting a fertiliser on to a growing crop. This is commonly done with nitrogen in the spring on winter cereals.
(d) *Contract spreading* is done by specialist firms which supply and spread fertilisers direct on the land, using large-scale spreaders. Agricultural salt and basic slag are often spread in this way.

Residual Values of Fertilisers

Lime and fertilisers are not all used up in the year they are applied to the land. Some of the value is left over for a year or more, and it helps crops grown later. When a farmer takes over a farm, he has to pay the outgoing farmer for this 'residual value'. The actual cash value is worked out by professional valuers, who use official tables of residual values.

CONTROL OF HEALTH OF LAND AND CROPS

Good farming keeps land and crops in healthy condition and grows the largest crops possible. Land and crops are kept in healthy condition by good cultivation and by keeping up the amount of organic matter, but also by the farmer's good management, by a sensible rotation, and by the control of weeds, pests and diseases.

Rotations

A rotation is a succession of crops usually fixed in a certain definite order. Rotations are not closely followed today as they were in the past, when it was often considered wicked to get away from a fixed rotation. In arable districts today as much corn is grown as possible, with other crops which give a rest from corn (break crops) used to break this sequence.

The main reasons for having any rotation are these:

1. To control pests and diseases of land and crops, such as eelworm in root crops and take-all in cereals.

2. To keep up or improve soil fertility, as by ploughing in leys.
3. To get the best yields from all crops.
4. To control weeds.
5. To cultivate the whole farm efficiently, avoiding cultivating to the same depth year after year.
6. To make sure manures and fertilisers are used to best advantage.
7. To spread the use of labour over the year.
8. Because it is required by terms of the contract with certain crops—such as sugar beet or seed crops.

To suit changing conditions and the needs of the market, any rotation needs to be flexible, so new crops can be introduced and changes made when and where needed. The opposite of a rotation is to grow one crop year after year. This is known as monoculture and can lead to troubles, particularly disease. The one crop which can be grown in this way is grass; permanent grass, if farmed properly, can stay productive for a long time.

Some Well-Known Rotations

The Three Field System—winter corn; spring corn; fallow—was followed in many parts of England during the Middle Ages until it broke down due to low yields, disease and new methods.

The Norfolk Four Course System—roots; barley; clover; wheat—was followed in the Eastern Counties during the eighteen and the early nineteen hundreds. As time went on, this rotation was extended and more corn crops included: roots; barley; barley; clover; wheat (sometimes barley next).

Ley Farming was followed in Scotland and in many parts of England, and is still practised. It needs a farming system where grazing livestock are kept: *Ley*; *ley*; *ley*; *potatoes*; *wheat*; *barley* or *oats* (undersown); or *ley*; *ley*; *ley*; *wheat*; *roots*; *barley* or *oats* (undersown).

Rotations today. It is best to look at rotations in your own district. Find out what rotation is followed (if any) by local farmers, and the reasons for the order in which the crops are grown. There is a much less rigid approach to rotations, and the rules are not followed so carefully as they were in the past. Modern methods, including varieties and chemicals, give a greater protection against the crop pests, diseases and weeds which originally provided the main reason for good rotations.

Today we find wheat following wheat, and other successions of cereal crops, with break crops in the rotation only when they are really needed. Probably the best way to consider a crop rotation is to record the percentage of cereals over a period of years. Thus a rotation used in the Eastern Counties: Wheat; wheat; peas; wheat; wheat; sugar beet can be expressed as 66 per cent cereals.

The Norfolk four-course system was 50 per cent cereals.

Weeds

The best time to kill most weeds is before you can see them, and the first way to kill them (some would still say the best way) is by cultivation. Good ploughing is still the basis of weed control in this country, but chemical methods become more and more efficient and sophisticated (and also more expensive).

The chief methods of weed control we have are these:

1. *Cultivation*

Ploughing is the first method and it is important to turn all rubbish in and cover it completely.

Stubble cleaning, using cultivators, makes many weed seeds germinate and the weeds are then killed by more cultivations or by ploughing—but leaving a stubble untouched is sometimes effective as the birds eat many of the weed seeds on the surface.

Harrowing in the early spring kills many seedlings.

Hoeing row-crops controls many weeds.

Fallowing, working bare arable land during the summer, gives a very good control of all weeds but is too expensive to be done much these days.

Burning stubble, so long as it is done with care, is effective and makes some weed seeds germinate earlier than they would do otherwise.

Rotary cultivation, gives a very good (although expensive) control of some weeds, even couch grass, if continued long enough—but if only done once or twice it may spread the weeds and make the problem worse.

2. *Rotation of Crops*

Each type of crop encourages its own set of weeds, so if we grow one crop, for example corn, too often, a number of weeds will be encouraged. By changing the cropping from one year to the next, many weeds are controlled. Some arable weeds will not grow in grassland nor some grassland weeds in arable.

3. *The Law*

You can be compelled by law to get rid of some weeds in your land. These 'injurious weeds' are: spear thistle, creeping thistle, curled dock, broad-leaved dock, ragwort. Also, seeds of the following weeds must be declared if present in seed samples: wild oats, docks and sorrels, blackgrass, couch grass, meadow grass, dodder.

4. *Chemicals*

Weedkillers (herbicides) are generally used on most crops, and every farm has a sprayer which can be used to apply these chemicals; some contract spraying is done for special purposes. There are many weedkillers in use, and more developed every year. They are sold under different names by the chemical firms, and this can be confusing. The two main types of weedkiller are:

(a) *Non selective*—these kill everything growing, and in some cases 'sterilise' the ground for a time.
(b) *Selective*—these kill only certain plants, or groups of plants, and so can be used to control weeds growing in most crops. They need to be used with care; dose rate, timing, soil conditions, temperature—all are important.

There are other words used to describe weedkillers, and it is as well to know what they mean. *Translocated* means that the chemical is taken into the plant and then moves about inside it. *Hormone* weedkillers are of this type and upset the growth of the plant, usually making it twist and distort, and often killing the underground parts of the plant, such as bindweed or thistles. *Residual* weedkillers persist in the soil and kill weeds as they grow or germinate, thus keeping the land clean of weeds for some time. *Contact* weedkillers scorch or kill nearly everything they touch, and thus can clear land or weeds completely, but do not always stop perennial weeds growing.

You must be careful with persistent chemicals in the soil or in crop produce. Sometimes a chemical lasts for some time and may affect a following crop; it may, for example, be in the straw of a cereal crop and affect some other crop such as tomatoes, if it is used for mulching.

Using Weedkillers

There are various ways in which these chemicals are used: it is important to know the various stages of crop growth, particularly the cereals, so that sprays are applied at the right time—see the chart on page 74.

1. Before the crop is sown or planted—*pre-sowing.*
2. After the crop is sown or planted, but before it comes through the ground—*pre-emergence.*
3. On the crop as it is growing—*post-emergence.*
4. Applied or worked into the land to kill perennial weeds—*residual weedkillers.*

Safety Precautions

1. Some weedkillers are poisonous and these must be treated with utmost care; you must know the regulations which control the use of poisonous chemicals.

2. Take care, particularly in windy weather, that sprays are not blown or sprayed onto neighbouring crops which may be harmed.
3. Do not overspray or double dose, as this will cause damage.
4. Follow the makers' advice on the use of any chemical.

Pests and Diseases

To control them properly, it is important to know something about the chief pests and diseases (see pages 44–45). But there is always something new and you may need to find out from a manufacturer or adviser just what to do for a particular pest or disease. These are the main means of control:

1. *Rotation*. Where any crop is grown year after year or too often, it can lead to trouble.

2. *Hygiene*. Take care of any place from which pests or diseases can spread. Remains of crops held over from one year to the next (old potato clamps), crop plants growing as weeds in the following crop, patches of waste land, heaps of rubbish, rough grass in ditches and dirty hedge bottoms, all these can cause trouble.

3. *Good Farming*. A crop which is growing well and strongly is less likely to be attacked than a weak crop. *Timeliness*—the ability to do tasks at the right time—is one of the most important qualities which contributes to good farming practice and which helps in dealing with pests, diseases and weeds.

4. *Chemicals*. Sprays and dusts may be used in two ways:

(a) Prevention, dealing with an attack before it happens. Seed of many crops is dressed to control fungus disease (wheat, barley, peas) or insect attack (wheat bulb fly, flea beetle on kale). Potatoes are sprayed to protect them against blight.

(b) Control. If a pest is found to be attacking a crop, it may be necessary to spray to give a control.

The chemicals commonly used are:

1. Fungicides, used to control fungus diseases.
2. Insecticides, used to control insects. Main types are: stomach poisons, contact poisons, systemic (which get inside the plant and poison insects feeding on it), granular (which are put into the soil).
3. Desiccants, used to burn or scorch off growing crops; in some cases to stop the spread of disease, as with potato blight.
4. Growth regulators, used to control the growth of certain crops, and in some cases to reduce the effects of heavy dressings of some fertilisers.

THINGS TO DO

1. Make a list of the main points about the district where you live, so far as they affect farming—type of soil, altitude, rainfall, temperature, prevailing winds, towns, markets, labour, etc.

2. Get and learn to use the ordnance survey maps which cover your part of the country; find your farm on all the different types of map.
3. Study the effect in the field of cultivating implements on different types of land.
4. Get and study advisory leaflets from your local office of the Ministry of Agriculture dealing with some of the practices and problems of farming in your area.
5. Look at the recommendations of more than one fertiliser manufacturer, and compare these for the main crops grown on your farm.

Make a plan of your farm and mark or colour in the different crops and grassland.

QUESTIONS

1. What are the best methods to judge the quality of land and the condition of a farm?
2. On land which needs draining, what methods can be used to get surplus water away from the soil?
3. How is soil sampled, and tested for acidity, and what scale is used to measure its acidity?
4. How are the plant food needs of a crop measured; what are all the separate plant foods which may be applied to crops; and what is the difference between straight and compound fertilisers.
5. What are the main methods of control of pests and diseases of farm crops? How are these carried out in practice?

* * * *

Make sure you study the leaflet, *The Safe Use of Poisonous Chemicals on the Farm* (MAFF).

Chapter 4

Cereal Crops

THE CEREALS—commonly called corn or grain—are the most important arable crops grown in Great Britain. The acreage of cereals has increased greatly in recent years; and with the help of fertilisers, weedkillers, fungicides, other chemicals and new varieties, it has become much easier to grow heavier yields of grain. Cereals are important cash crops when the grain is sold off the farm. A good deal of grain—mostly barley and oats—is kept on the farm for feeding to livestock, and straw is used for feeding, bedding and processing.

There have been several big changes in cereal growing during the last ten years, apart from the varieties—which are changing all the time and making higher yields possible. The *combine harvester*, used even on small farms and in districts which are only partly arable, has taken the hard work out of the harvest and reduced the need for labour at this time of the year. The arable peak period now comes *after* the grain harvest on many farms, into the Autumn, with straw disposal, ploughing and cultivations.

Selective weedkillers are simple to use with cereals, so that today these also are 'cleaning crops'. In the past cereal crops were much troubled by weeds, and needed a rotation which included cleaning crops such as roots, to keep the land clean.

Rotations have changed, and much more corn is being grown on many farms. At one time it was difficult to grow cereal crops too often in the rotation because of take-all, eyespot and other diseases, and it was considered bad farming to grow corn too often in any field. This is no longer the case, and by the use of an occasional 'break crop' (something to give a rest from corn) yields of healthy cereal crops are now grown year after year.

The main trouble now with the cereal crops, particularly if they are grown for a number of years in succession, is the build-up of weeds; the worst of these are the 'grass weeds', which include wild oats, blackgrass, onion couch, meadow grasses (poa) and couch grass. The exact problem may vary from one area of the country to another, but this group of weeds causes trouble in most of the arable areas, and forces the farmer to spend heavily on the right weedkillers.

The main cereal crops are wheat, barley and oats, with some rye grown on very light soils, and a small amount of mixed corn. In Great Britain maize is not often grown for its grain, although some work is being done on new varieties; it is grown as a forage crop (see page 116).

It is important to be able to recognise the different cereals when they are growing. With practice, you can tell one crop from another by leaf colour and general appearance, but to check on this examine a few plants. The diagram below will show you how the cereals vary. Look for the auricles (claws at the point where the leaf meets the stem) and remember:

Wheat has whiskers on the auricle; barley has big, bare auricles; rye has very small auricles; and oats have none.

In the past, yields and seed rates were given in capacity measures (bushels, sacks and quarters); in more recent times weights have been generally used.

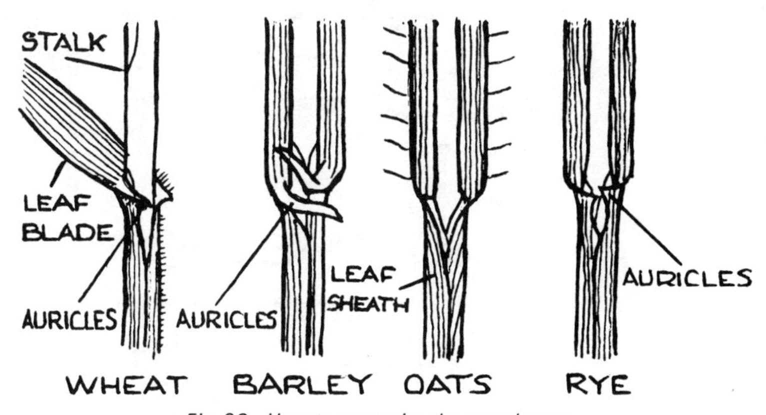

Fig. 20. How to recognise the cereal crops.

Methods of sowing cereals have varied over the years; the combine drill which puts fertiliser into the soil with the seed is still preferred on some farms, and other methods of drilling or spreading seed and fertiliser separately are in common use, particularly on large arable farms.

Conventional methods involve ploughing, cultivating and then drilling the seed. There are methods which cut out the use of the plough, instead making greater use of the cultivator; this is 'minimal cultivation'. There is also 'direct drilling', by which the seed is put direct into stubble or undisturbed land with special machinery.

A modern practice is to aim for fairly precise 'seed populations' with a known number of seeds or weight of seed per acre or hectare. There are various 'blueprint methods' of growing heavy crops of cereals, with the various jobs (drilling, fertilising, spraying) carried out at precise times according to the growth of the crop.

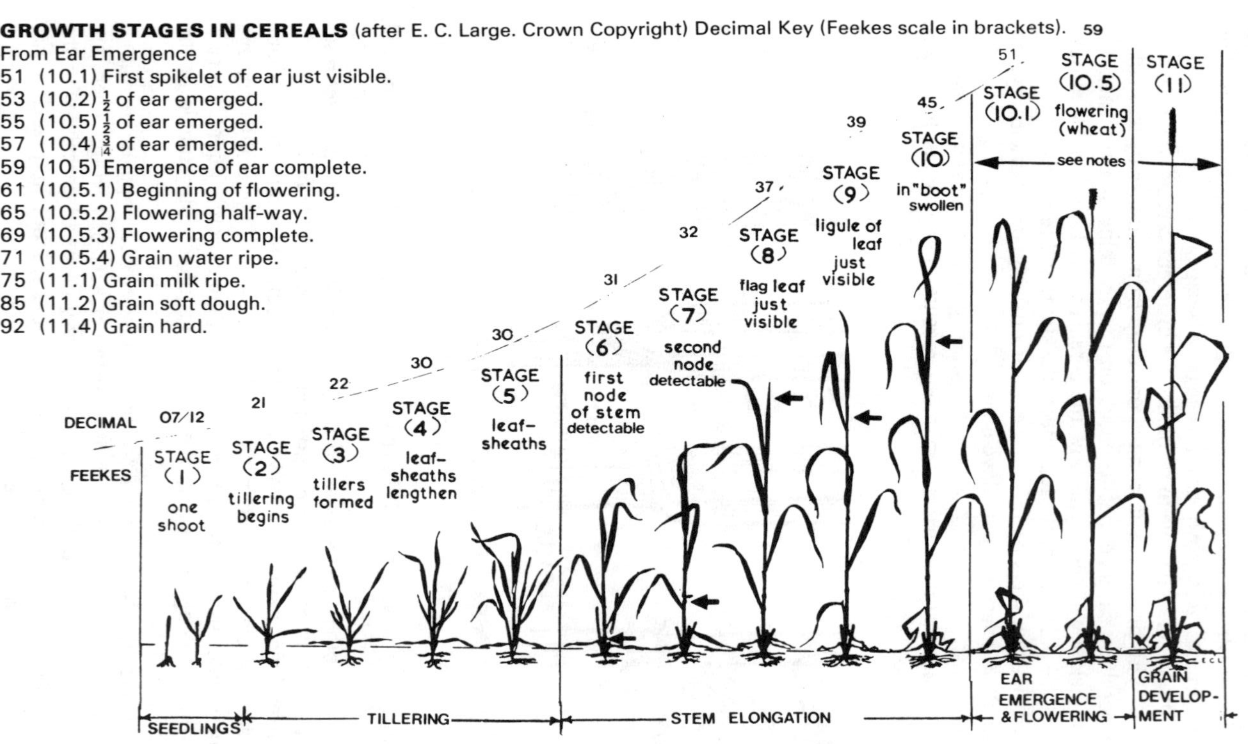

Fig. 21. Stages in the growth of cereal crops.

The 'tramline technique' is practised by some cereal growers today. It means leaving specially measured wheel tracks through the crop at the time of drilling, so that all the later tractor operations—fertilising, spraying etc.— can be carried out along the same wheel marks with the least possible damage and disturbance to the crop.

It is important to know the various stages of growth of cereals, so that crops can be sprayed at the right time, and other jobs timed just right for the benefit of the crop. The diagram opposite shows the stages of cereal crop growth.

WHEAT

Wheat is the cereal crop which pays the farmer best, so long as he can get good yields. It is a cash crop, and is usually sold off the farm; although wheat can be used for feeding to livestock, it is not so popular nor so widely used in this way as barley.

The best conditions for wheat are found in the eastern and southern counties; the crop should be able to mature and be harvested in good condition, which makes it difficult to grow in wet and late districts.

In the past, wheat was considered a crop for medium and heavy land in a good state of fertility. Today, with modern fertilisers and cultivation methods, bigger yields can be grown with less trouble on a wider range of soils, but there is no point in growing wheat on land which is too poor or light.

In the past it was usual to give wheat a place in the rotation which encouraged higher yields—for example, after a ploughed-up ley, after beans or peas, or a well-manured root crop. Today it is not unusual to find first, second and even third wheat crops grown one after the other. This is alright so long as care is taken with choice of varieties and disease control.

Types: There are winter (autumn) wheats and spring wheats, with different varieties in each type. Which is used depends upon the farming system, previous cropping and local conditions. Winter wheat yields about 15 to 20 per cent more than spring wheat.

Ordinary varieties of wheat produce grain which is used for animal feeding. There are certain varieties which, when well grown, produce grain of superior quality used for milling into flour for human use—as bread, biscuits etc. These are known as 'milling wheats' or sometimes 'quality wheats', and make a better price when sold to the grain merchant.

Another special type of wheat which is grown in Britain is the 'durum wheat' used for the production of pasta foods—spaghetti, macaroni, etc.

Seed: Seed wheat is usually bought in; occasionally wheat grown on the farm is used. It is usual to have seed dressed with chemical seed dressings against fungus disease and also against certain insect pests (if trouble is expected).

Sowing: Ploughing for all cereal crops is shallow, 10–20 cm (4–8 in). Wheat is usually drilled with a row width of 18 cm (7 in), sometimes narrower at 10 cm (4 in) rows; it can also be broadcast. The best depth is 2·5–4 cm (1–1½ in). Traditionally an ideal seedbed for winter wheat is firm but fairly rough and cloddy on top; but today some good growers look for a finer seedbed so that chemicals for weed control can be applied properly. A seedbed for spring wheat should be in any case finer. It is not always necessary to plough for a wheat seedbed. The land may be prepared by other cultivations (chisel-ploughing or discing) or the seed may even be sown direct into the surface. Winter wheat is best drilled in September/November, at 150–185 kg per hectare (60–75 kg per acre). Spring wheat is drilled February/March, at 185–250 kg per hectare (75–100 kg per acre).

Fertilisers: Unless the land is well supplied with plant foods, it is usual to give winter wheat some P and K in the seedbed, sometimes with a small amount of N as well. Most of the N is given in the spring, as a 'top dressing'.

(Refer to page 65 for information about normal rates of fertilisers.)

For spring wheat, all the fertilisers should be applied in the seedbed. The land must not be short of lime.

Spring Work: Wheat is sometimes harrowed in the spring, to settle the soil down and to kill weed seedlings, but many good growers do not harrow at all. Sometimes this crop is rolled, and this does put stones into the soil and out of the way of the combine later.

In the past, wheat which has made good strong growth over the winter could be grazed in the spring, but this is seldom done today.

Small seeds (grass, clover) may be undersown in wheat in March/April, and harrowed in lightly.

Spraying: The wheat crop is usually sprayed for weed control—both grass weeds (see page 120) and broad-leaved weeds being common. The use of fungicides and of growth regulators is also common.

Harvesting: Wheat is harvested during August and September, winter wheat being ready before spring wheat. Most wheat is cut with a combine harvester. It must be in dead ripe condition (the grain hard and dry); if it has to wait too long, it will start to 'shatter' and some of the grain will be lost on the ground.

If wheat is needed for a special purpose, such as the supply of thatching straw, it may be cut with a binder 7–10 days earlier than it would be with the combine.

Combined straw is usually baled for use on the farm or for sale. It may also be ploughed into the land, either long or chopped; or it may be burned, taking suitable precautions. There is a National Farmers' Union (NFU) Code of Practice for straw burning which should be followed carefully.

Storage. Wheat, like other grain, can be stored loose (in heaps), in sacks, or in bins and silos. When harvested, wheat may have a moisture content of 18 per cent or more. Safe levels for storage of wheat and other grain are:

In bulk (long time)	Under 14 per cent moisture content
In bulk (temporary)	14 to 16 per cent moisture content.
In sacks (long time)	16 to 18 per cent moisture content
In sacks (temporary)	18 to 20 per cent moisture content.

Wheat which is to be dried should first be put through a cleaner, to remove rubbish, dust, etc. A lower drying temperature is needed for seed wheat (up to 48°C: 120°F) and for flour milling (up to 65°C: 150°F). For feeding on the farm a higher temperature is suitable (up to 82°C: 180°F).

Yields

	Average	*Good Yield*	*Our Farm*
Grain—Winter			
Tonnes per acre	2·2	2·6	
Tonnes per hectare	5·5	6·5	
Grain—Spring			
Tonnes per acre	1·7	1·9	
Tonnes per hectare	4·2	4·8	
Straw			
Tonnes per acre	1·0		
Tonnes per hectare	2·5		

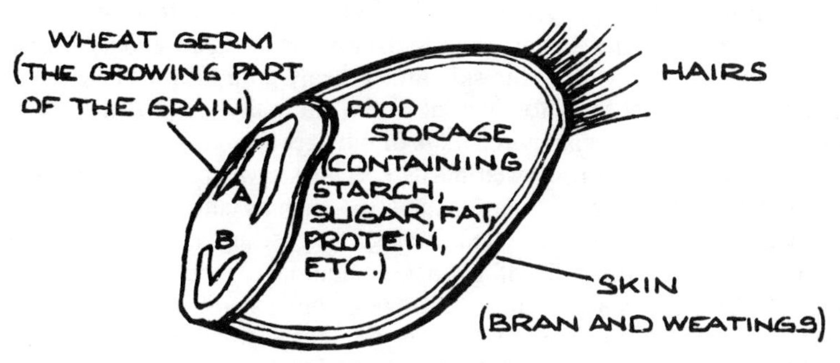

Fig. 22. A grain of wheat.

Uses

Wheat is normally fed to livestock—whole (for poultry) and ground, crushed, or rolled (for pigs, cattle and sheep).

For human use it is made into bread, flour, and biscuits; a small amount is used for distilling. The 'wheat offals' produced as a by-product of milling include bran and middlings (weatings) and are used for animal feeding.

Wheat straw is of no use for feeding; it is commonly used for littering livestock, and may be used for thatching.

BARLEY

Barley does not need such good conditions as wheat and can be grown on a wider range of soils. It is looked on as the best of the cereals for animal feeding—it can be fed to all types of livestock—and for this purpose it has partly taken the place of oats.

Although it is commonly kept on the farm for feeding, barley is also a useful cash crop. The best conditions for barley are found in the chief arable areas of the south and east, but it is grown in most parts of Great Britain.

Malting Barley

Malting is a process of treating grain which is mainly done with barley; in Britain about 20 per cent of the barley crop is used for malting. Malting barley is carefully selected, from suitable varieties (consult the NIAB lists), mainly grown on light to medium soils, drilled early into good seedbeds and kept free from fungus diseases and arable weeds. It must be harvested carefully, protected against damage and dried to a moisture content of 15 per cent or less, and then stored properly.

On suitable land (not too wet or heavy) it is possible to grow malting barley. This fetches a better price, but must be of good quality—clean, dry, well harvested, with a thin skin and a mealy grain. Barley is tested for its nitrogen content, and for malting is not normally taken if the the test shows more than a certain percentage of nitrogen.

To grow malting barley well needs special knowledge and skill. Some districts also have a reputation for quality—such as land near the sea. The first winter barley crops harvested often go for malting.

Malting barley must be of good germination (95 per cent or more), moisture content not too high, nitrogen content not too high (up to 1·85 per cent), a good, clean sample, free from small grains and weed seeds. Check the current standards with your local barley buyers.

Malting barley is bought by maltsters, who process it in buildings known as maltings. In large containers, it is soaked in water and allowed to germinate under controlled conditions so that the young roots and shoots within the grain start to develop. After about a week of growth, the grain—in which the stored starch was turned to a form of sugar under the influence of chemicals known as enzymes—is heated and dried to stop the process of growth. The grain is now known as malt, and this is used for brewing to produce beer and vinegar, for the making of malt products, and for distilling to produce spirits.

Feeding Barley

This does not need such special conditions, and can be grown on most soils. More fertiliser (particularly nitrogen) can be used to push up yields as high as possible, and may increase the amount of protein in the grain, which is useful. Most barley is grown for feeding.

Types: There are both winter (autumn) and spring barleys, with several varieties of each type; winter barley gives a higher yield. The amount of winter barley has increased greatly in recent years, at the expense of spring barley.

Seed: Seed barley may be bought-in or home grown (so long as it is clean, and there is no mixture of varieties). It can be dressed with chemicals against disease (smut, leaf stripe) or pests (wireworm).

Sowing: Barley needs a fine seedbed—finer than for any other cereal. The crop may be broadcast, but is usually drilled in rows 18 cm (7 in) apart, with a seed rate of 125–185 kg per hectare (50–75 kg per acre) and 2·5–4 cm (1–1½ in deep), February/March. Winter barley is drilled September/November at 150–185 kg per hectare (60–75 kg per acre).

Fertilisers: The full dressing of fertilisers is best applied in the seedbed (except in the case of winter barley, which is treated like winter wheat).

(Refer to page 65 for information about normal rates of fertilisers.)

The soil must be well supplied with lime, or barley will fail.

Spring Work: The crop can be harrowed, but for malting barley this must be done with care. Spraying against weeds must be done with care on malting barley, as it may affect the grain. The crop is often sprayed against leaf diseases. Barley may be undersown with grass and clover seeds.

Harvesting: Barley is harvested with the combine harvester. It should be dead ripe, after the ears have 'necked' (turned downwards). Malting barley is left longer in the field than feeding barley to make sure it is really mature, which improves its quality.

Storage: Safe moisture levels for storage are the same as those given for wheat. When barley is to be dried, this must be done with care, particularly if the grain will be used for malting. Malting and seed barley must be dried at a lower temperature (up to 48°C: 120°F) than feeding barley (up to 82°C: 180°F) and it must also be dried more slowly.

Yields

	Average	*Good Yield*	*Our Farm*
Grain—Winter			
Tonnes per acre	1·9	2·2	
Tonnes per hectare	4·8	5·5	
Grain—Spring			
Tonnes per acre	1·7	1·9	
Tonnes per hectare	4·2	4·8	
Straw			
Tonnes per acre	0·8		
Tonnes per hectare	2·0		

Uses

Most of the British barley crop is fed to cattle, sheep, pigs and poultry (usually ground or rolled). The rest is used for malting or distilling, and a small amount for human food. Barley straw is used for feeding to cattle, but is of low value. It forms useful litter, more absorbent than wheat straw, and softer.

OATS

Oats are grown in most parts of this country, but are found more in the wetter northern and western districts where wheat and barley are not so successful. Oats can also be grown on land which is moderately acid, where barley or wheat would fail. Both the grain and straw from this crop are used mainly for feeding on the farm.

In recent years less oats have been grown, particularly in the Eastern Counties; barley has taken the place of oats for feeding.

Oats are grown much like other cereals, but two points are important. They must not be grown too often on any land, as this can cause trouble with eelworm (a soil pest). If grown on land which is too rich, or if fertilised too heavily with N, oats will not stand—oat straw is weaker than that of other cereals.

Types

There are both winter (autumn) and spring oats, with different varieties in each type. Winter oats give a higher yield, and are less likely to be damaged by the frit fly pest. Frost damage can occur in a hard winter.

Seed

Oat seed needs to be harvested well to produce maximum yields. It is usually dressed with chemicals against disease (smut, leaf-stripe) and pests (wireworm).

Sowing

Oat seed is thinner than that of other cereals, and is covered with a husk. It therefore weighs less, and the seed rate must be adjusted to make sure of enough plants per acre.

The crop may be broadcast, but is usually drilled in 18 cm rows, 4–5 cm deep, at 185–250 kg per hectare (75–100 kg per acre).

The seedbed should be similar to that needed for other cereals. Winter oats are best drilled September/October, and spring oats February/March.

Fertilisers

Fertilisers are given to oats in the same way as wheat, but with less nitrogen.

(Refer to page 65 for information about normal rates of fertilisers.)

Lime is not usually needed. On soils with plenty of lime, a shortage of manganese can cause trouble, and this is becoming more common.

Spring Work

Oats are treated much like other cereal crops. A well-grown crop of winter oats may be grazed up to mid-April, but this may reduce yields, and it is important to give a top-dressing of N.

Harvesting

Oats are combined just before fully ripe; if left too long, they very easily 'shatter' and corn is lost on the ground.

Storage

Safe moisture levels for storage are the same as those given for wheat.

Yields

	Average	*Good Yield*	*Our Farm*
Grain—Winter			
Tonnes per acre	1·9	2·2	
Tonnes per hectare	4·8	5·5	
Grain—Spring			
Tonnes per acre	1·7	1·9	
Tonnes per hectare	4·2	4·8	
Straw			
Tonnes per acre	0·8		
Tonnes per hectare	2·0		

Uses

Most oats grown in this country are kept on the farm for feeding, either rolled or ground, to cattle, sheep and horses. Finely ground, they are sometimes used for pigs and poultry.

A small amount is sold for human use, mostly as oatmeal.

Oat straw is used for cattle and sheep, and has a similar feeding value to barley straw.

RYE

In this country, rye is not much grown for grain production—as a rule it is found only on poor, light, acid soils, where it will stand up to conditions that prevent other cereals from growing.

It is very hardy and can stand cold, hard, winter conditions better than any other cereal crop.

It is useful as a green crop (see page 115) for early grazing in the spring.

Types

Although there are both winter and spring types, only the winter type is grown for grain production in this country; a few varieties only are used.

For grazing, there are special varieties.

Seed

Rye seed is usually bought-in fresh each time it is needed. It is a crop which cross-fertilisers easily, and varieties cannot easily be kept pure.

Sowing

This crop is treated just like winter wheat. It may be broadcast or drilled at 100–250 kg per hectare (75–100 kg per acre) in September/October.

Fertilisers

If rye is given too much N it will not stand. Normal dressings as for other cereals, but with less N.

Harvesting

Rye is the earliest of the cereal crops to be ready for harvest, and should be cut when it is dead ripe. Rye has long straw which is sometimes difficult to harvest and slows down the speed of the combine.

Yields

As rye is usually grown on poor land, the yields are not high. Average is about 3·0 tonnes per hectare (1·2 tonnes per acre). with 3·7 tonnes of straw per hectare (1·5 tonnes per acre).

It is suggested now that good yields of rye can be obtained on better land, with the use of growth regulators to make sure the crop stands.

Uses

Rye can be used for stock-feeding like other cereals, but it is not commonly used. Most of the rye harvested goes for seed for forage crops. Some is used for human consumption as rye crispbread.

Rye straw is useless for feeding. It can be used for thatching.

BREAK CROPS

It is sometimes possible to grow cereal crops for a number of years one after the other. If it goes on for a long period, it is known as continuous cereal growing. In the past, this practice often led to trouble with weeds, diseases and pests. Today, there is better control of these by chemicals, particularly fungicides and weedkillers which deal with grass weeds, and thus farmers go in for longer runs of cereal crops.

In practice, it is common to grow wheat for two or three years, barley for longer, and then to break the succession with what are called ‘break crops’ or sometimes ‘change crops’.

Farmers need to grow break crops which are at least as profitable to grow as the cereal crops they displace. The choice of break crops varies with costs, prices received, and how they fit into the farming system. There are three main types of break crop:

(a) Combinable break crops—peas, beans, oil seed crops and other seed crops. Two cereals—oats and maize—are 'partial break crops' which help to halt the increase of some diseases.
(b) Other break crops—root crops and forage crops.
(c) Grassland.

THINGS TO DO

1. See NIAB leaflets and seed catalogues from local and national seedsmen, to find out what cereal varieties are on sale.
2. See samples of grain of different quality so that you can tell good quality from poor quality.
3. Make a list of varieties of winter wheat and spring barley or oats grown on your farm or another local farm, or the College farm.
4. Use a moisture meter to test samples of grain.
5. Check on local prices for seed and for grain and straw of the main cereals grown in your area.

QUESTIONS

1. What percentage of cereal crops are grown on your farm, or a typical local farm? Give a separate figure for winter and spring cereals.
2. How do you tell the difference between the various cereal crops in May?
3. What methods are used to dispose of straw on your farm or in your area?
4. What are average yields of cereal grain and straw on your farm and in your area?
5. What is the seed rate, row width, and time of sowing of the main cereal crops grown on your farm?

Chapter 5

Combinable Break Crops

BEANS, PEAS AND SOME OTHER CROPS

LARGE ACREAGES of beans and peas were once grown for livestock feeding—these crops produced valuable protein-rich grain which could be mixed with home-grown corn for cattle, sheep, and also pigs. These crops belong to the legume family, and through the bacteria in the roots they take nitrogen from the air and feed themselves and the soil with this important plant food.

On heavy clay soils, beans were used on up to one-quarter of the arable acreage, and peas took their place in the south and east on lighter and drier soils. In recent years growing both beans and peas for stock feeding has become much less profitable compared with cereals.

These crops have not been dropped altogether, for different types are being grown today and acreages have increased, particularly of peas. The types grown now are for human use. With the development of canning, deep freezing and accelerated-freeze drying, there is a growing market for green peas—harvested as fresh and as quickly as possible—and for green beans, both grown as field crops under carefully controlled conditions, usually on contract.

There is also a large acreage of peas for harvesting, combined and treated much like grain.

Apart from their cash value, some of the importance of all the crops included in this section of the book is in their use as break crops (change crops), which give the land a rest from cereals. Flax, linseed and lupins are among the crops which are sometimes grown, but more valuable cash crops at the moment are oil seed rape and various seed crops. As prices and demands alter, so certain crops become more or less attractive to the farmer. For this reason, and because of technical changes in growing methods, the use of break crops on the farm may change quickly from one year to the next.

FIELD BEANS

Field beans were once an important arable crop—mostly on heavy soil, which was often called 'wheat and bean land'. The importance of this crop was for the production of home-grown protein food for livestock and in adding nitrogen to the soil for the next crop. For a number of reasons, including weeds, disease (chocolate spot) and pests (aphis, weevils), and because there have been few good varieties or strains of beans, this crop has been grown much less in recent years.

Things have improved recently. Spraying can deal with pests and diseases, there are weedkillers which can be used on bean crops, and new varieties of beans have been developed; so that it is possible to grow good crops with much less trouble.

Beans do best on the heavier soils, although spring beans are sometimes grown on lighter land, but the land must be well drained and not acid. This crop cannot stand very hard winters, and so some districts are not suitable.

Types

There are separate types of field beans, with a few varieties of each:

Winter (autumn) beans.
Spring beans: (a) Horse beans—with seeds like winter beans.
(b) Tic beans—with smaller seeds.

Because of weather conditions, winter beans are not usually grown in the north, where spring beans are grown instead. The winter crops give a heavier yield, by up to 20 per cent.

Sowing

Beans need a seedbed similar to that suitable for cereals, although it may be rather deeper—cloddy on the top for winter sowing, and finer in spring.

Winter beans are best drilled in the first half of October at 250 kg per hectare (100 kg per acre). Spring beans are drilled in late February or up to mid-March at the same rate (slightly more for horse beans). Seed of winter beans can also be ploughed in.

Seed is put in 5–7 cm deep, and can be drilled either in wide rows of 50 cm or more, at 30–35 cm or in 18 cm rows like the cereal crops.

Fertilisers

The land must not be short of lime. Dung gives good results with this crop. Refer to page 65 for information about normal rates of fertilisers. (Beans suffer if there is a shortage of K in the soil.)

Spring Work

Winter beans are sometimes harrowed in the spring, like cereal crops. Weed-killers are commonly used; crops grown in wide rows can be hoed.

Harvesting

Beans are harvested with the combine, and the crop is left until the leaves have withered and most of the pods are dry and brittle.

Spring beans are late to harvest, and may not be ready until September/October.

Storage

Beans can be stored like grain. Drying must be done slowly and at a low temperature.

Yields

	Average	*Good Yield*	*Our Farm*
Grain			
Tonnes per acre	1·1	1·3	
Tonnes per hectare	2·7	3·2	

Uses

Both winter and spring beans are sold as a cash crop, and often go for export. Beans may also be used on the farm for feeding to all types of livestock, either ground or cracked. Beans are better fed after a period of storage or drying. Tic beans are dealt with in the same way, but are often sold for feeding to pigeons. Combined bean straw is hard, brittle and of little or no feeding value.

GREEN BEANS

This type of bean, which has developed from a garden crop also known as French bean or dwarf bean, is now grown as a large-scale farm crop. Much of it is grown by individual farms and farming groups as a contract crop for processing. It is a fast-growing productive crop, needing good land and good soil conditions, which is mainly grown in East Anglia and Lincolnshire.

This crop needs specialised and expensive equipment and a great deal of technical knowledge. Seed is sown in May/June into a good seedbed so that the crop can grow without check. It is harvested by large viners in the field and sent off direct for processing, to produce frozen, canned or freeze-dried beans for sale.

OTHER BEANS

Broad beans and runner beans (both bush and climbing types) are sometimes grown on farms as vegetable crops, and are also grown as seed crops. There has also been some interest in other beans: navy beans grown to produce seed for the canned baked bean trade, haricot beans which are sold dry in the grocery trade, and soya beans which are not yet successful in this country but may have a future as a bean for industrial processing.

PEAS

Peas are grown mostly in the east of this country, which gives the warm sunny conditions they need. Peas for livestock feeding are seldom seen now; most are grown for human consumption. There are three main types:

1. *Vining peas* which are cut green and processed by a viner to provide peas for canning, quick freezing or dehydration.
2. *Fresh peas* which are picked by hand and are now only grown on a very small scale.
3. *Dried peas* which are combined in the field and used for processing, for seed, or for sale as dried peas.

The buyers of green peas for vining control the growing of the crop—times of sowing and harvesting—and insist on high quality produce.

In the past, peas were always considered a weedy crop which made the land dirty. Sprays for weed control now take care of this problem, and sprays also deal with the insect pests which attack this crop.

Sowing

Peas need a seedbed similar to that for spring cereals. They can be sown from as early as January (or even December in sheltered places) until April or May. Peas for harvesting dry are usually sown in early March; peas for picking or harvesting green are sown to make the best prices on the market and in some cases to have a succession of crops so there is a steady supply.

Seed is drilled at about 5 cm deep, 250 kg per hectare (100 kg per acre), in rows 18 cm apart or wider.

Fertilisers

The land must not be short of lime, or peas will fail.

Refer to page 65 for information about normal rates of fertilisers.

Spring Work

Peas are a dirty crop, and weeds can be troublesome. Several types of weedkiller can be used with peas, but spraying must be done with great care, and timing is very important. Peas are also sprayed against important pests, such as aphis and pea moth.

Harvesting

Dry: Peas are ready to cut when the stems (haulm) are yellow and the pods have turned colour. If left too long, the pods will shatter and some of the peas will be lost on the ground.

The crop may be sprayed with a chemical desiccant to ripen it and make it ready for harvesting. After this, it can be direct combined from the standing crop, or cut with a pea cutter into windrows (swaths) and then combined from the windrow. With good harvesting weather, the crop can be cut without desiccation.

Tripods and four poles have been used for peas to allow the crop to dry properly before combining or threshing, but this method needs plenty of labour.

Green: Some peas are still grown for hand-picking; this needs a big gang of pickers and the crop must be cleared in good condition as quickly as possible.

When peas are being grown for vining, the crop is cut in the field by mower or pea-cutter and then put through the viner or through the pea podder.

Storage

Harvested peas and beans can be stored and dried but this must be done carefully.

Safe levels of moisture content for storage are:

In bulk (short time)	17 per cent
In bulk (long time)	14 per cent
In sacks (short time)	18 per cent
In sacks (long time)	15 per cent

Peas must be dried slowly and to a lower temperature for seed 38°C: 100°F than for human use 43°C: 110°F.

Yields

	Average	*Good Yield*	*Our Farm*
Seed			
Tonnes per acre	1·25	1·4	
Tonnes per hectare	3·1	3·5	

Uses

Peas harvested for livestock feeding can be used in the same way as beans.

Maple peas are sold for feeding to pigeons.

Peas harvested dry for human use are either sold in packets or loose for cooking, or are soaked and cooked in the factory and sold in tins as 'processed peas'.

Peas picked green are sold in the pod through greengrocers' shops. Vined peas are used for canning ('garden peas') for deep-freezing or for dehydration.

Green pea haulm from the viner is a good bulky food enjoyed by livestock, and can be fed fresh or made into silage for feeding later. Dry pea straw is a good food for cattle and sheep but after chemical desiccation the straw is useless.

OIL SEED CROPS

A number of crops are grown, in Britain and in other countries, for the production of seed which is rich in oil. The oil is extracted from the seed, and is used for a number of industrial purposes—including cooking oil, margarine, paints, other chemical uses, and even in the past for fuel. The oil is extracted by crushing and by other methods, and the material left over after this is done is called 'oil cake'; some of it is used for animal feeding (see *Farm Livestock* book).

Crops used for the production of oil include oil seed rape, linseed, sunflower, and occasionally poppy seed and mustard seed, all of which have been grown in Britain; and in other countries soya bean, palm kernel, cotton seed, groundnut and maize. At the present time, oil seed rape is the only crop of this type grown to any extent in Great Britain.

OIL SEED RAPE

There are both winter and spring types. Winter rape is sown from mid-August to mid-September and harvested in late July. It can be damaged by pigeons during the winter. It gives a higher yield than the spring type which is very little grown now. Spring rape is drilled in mid-March and harvested in September.

Sowing

The seed needs a fine, firm seedbed; it may be direct-drilled into a stubble. The seed is treated aginst flea beetle damage to the developing plant. It may be broadcast or drilled in narrow rows, at 7 to 9 kg per hectare (2·8 to 3·6 kg per acre).

Fertilisers

The soil should not be acid.

(Refer to page 65 for information about normal rates of fertilisers.)

Spraying

Various weedkillers can be used but it is not easy to control all weeds in this crop. Spraying may be necessary against insect pests, including beetles and weevils.

Harvesting

The crop is sometimes swathed and then combined from the swath, or may be direct combined after chemical desiccation. The seed is very small and can easily leak away through holes and cracks.

Storage

The seed contains more than 40 per cent valuable oil and must not be allowed to heat. If it is to be stored for more than a short time, it needs to be dried down to 8 per cent moisture content.

Yields

Winter rape can produce 2·5 tonnes per hectare (1 tonne per acre); spring rape 2 tonnes per hectare (0·8 tonnes per acre).

LINSEED

The same plant produces both linseed, which is an oily seed, and flax, which is a strong fibre found in the stem. There are several different varieties of each type. Both crops were grown in this country but there has been little interest in either of them for some years.

This crop needs the same sort of treatment as spring corn, but the seed-bed should be finer and rather firmer, as the seed is small; and clean because the small plants do not compete very well with weeds.

The seed is drilled, or broadcast, in March/April at about 60 kg per hectare (25 kg per acre) for linseed (or twice this rate for flax production).

Seed Production: Small shiny brown seeds are produced in 'bolls' at the top of the stem. When the seeds have coloured and filled out, the crop is cut like corn. Average yield is 1·8 tonnes per hectare (0·75 tonnes per acre).

Linseed is used for the production of linseed oil, which is pressed from the seeds and sold for paint making and other purposes. What is left is called linseed cake—a valuable protein food for livestock feeding. Whole linseed is sometimes used as a tonic or medicine for animals.

Linseed straw is no use for feeding, and too hard and non-absorbent for bedding.

Fibre Production: It is important not to break the long fibres which run up the stem, and for this reason the crop is pulled out of the ground rather than cut, while it is still supple and not too dry, using flax-pulling machines. The fibre is taken from flax stems by factory processes, some of which involve soaking in water. This fibre is made into linen.

SEED PRODUCTION

Seed Selection

The farmer needs seed which is:

Pure —free from mixture with other strains or varieties.
Clean —free from weed seeds or rubbish.
Healthy —free from disease.
Vigorous —quick growth, leading to better and more even establishment in the field.
Good Germination—a large proportion of the seeds will grow.

Seed Quality

This is important to the farmer and to every other user of seed. With the exception of potatoes, the cost of seed necessary to grow any crop is only a small proportion of the crop's value.

The law gives some protection to anyone buying seed. The buyer must be told if cereal, grass or clover seed contains any of the following weed seeds: wild oats, dodder, docks, wild radish, black grass, couch grass.

Certain vegetable seeds must be tested by law, and details of the purity and the percentage of germination must be given to the buyer if they are below certain levels.

Anyone can have samples of seed tested by the official seed-testing station for a small charge; this testing gives reliable information about purity, germination, and diseases which can be carried with the seed.

Seed Crop Inspection

To cover purity, right varieties and freedom from some diseases, seed crop inspection is done for some crops in co-operation with NIAB (see Introduction). These crops include many types and varieties of cereals, grasses, clover and vegetable seeds.

Seed potato crops are inspected by the official agricultural departments, and sold on grade according to purity and freedom from virus disease.

In some counties, there are local seed production associations, by which growers have their crops inspected, handled, and sold under special names.

SEED CROPS

Cereals for seed are grown in just the same way as crops for sale or for use on the farm. The grower must be careful to use seed of high quality, and to sow it on clean land—which does not contain seed of other cereal crops or other varieties of the same crop, or the harmful weeds. Seed crops must be harvested in good condition, and dried, cleaned, and handled with great care.

Beans and peas are grown in the same way as normal crops, but treated with the same care as cereal seeds to make sure they are clean and healthy.

Potatoes are grown from tubers, not from seed, and are usually produced from crops which are harvested early. Seed of the highest quality comes only from certain parts of the British Isles.

Sugar beet and mangels are biennial plants, which produce a seed crop in the second year of growth. The seed crops are grown mostly in the Eastern Counties, and there are zoning schemes to make sure that crops of this family do not cross-fertilise with one another (sugar beet, mangels, red beet, and some other garden crops).

Seed is sown in narrow rows during the summer, and the small plants (known as stecklings) are transplanted that autumn or the following spring, and put out in wide rows with plenty of room for the plants to develop. The seed crops are cut by hand, tied up, and stooked in the following late summer.

Turnips, swedes, kales and rapes are also biennial crops, harvested in the year after they are sown. The seed is sown in wide rows in late summer, and the plants thinned out the following spring (swedes are sometimes transplanted during early spring). These crops are cut, put into a windrow and combined.

Grasses are grown for seed in several parts of this country, and all the common grasses are grown in this way. Seed is sown thinly, and the crop established in the first year (either in wide rows, or with the ryegrasses in narrow rows or broadcast) and harvested the following year. The seed crop can be cut and combined from the swath or direct combined. Most of the grass seed crops are harvested for two years in succession.

Clovers are drilled or sown broadcast early in one year, often under a cover crop of cereal, and cut for seed the following year. It is common now to cut and then combine from the row afterwards. Red clover can be treated with a chemical desiccant to kill the leaf and then combined direct.

Red clovers either give one seed harvest (broad red) or sometimes two (late flowering red) and white clovers are cut several years in succession.

With many seed crops, isolation is important, so that there is no cross-breeding between types or varieties. The necessary distances vary and can be found in information from N.I.A.B.

THINGS TO DO

1. Compare the different types of peas which are sold for human use—and how they are sold.
2. Inspect samples of pea seed and see the points which determine quality.
3. Inspect growing crops of field beans and look for spacing between the rows, and for signs of pests and diseases.
4. Find out the price paid for oil seed rape or other seed crops and relate this to average yields of such crops.
5. Study the conditions laid down in contracts for production of seed crops, and find out about crop inspections.

QUESTIONS

1. What difference in yield would you expect between winter-sown and spring-sown field beans?
2. What particular care needs to be taken with rape seed in store, and why is this?
3. What methods are used to harvest pea crops—both dried and green peas?
4. What are the main points about seed quality, as it affects the farmer?
5. Which seed crops need to be grown separately from other crops, and why is this?

Introduction to Root Crops and Forage Crops

The forage crops and the root crops—including potatoes, which are really tubers—produce the heaviest yields of any crops grown in this country. Roots and green crops have a high water content, but even taking this into account, the amount of food produced is high compared with cereals, or other combinable crops.

Roots are expensive to grow, needing five to ten times as much labour as cereal crops—but if they are well grown the cash root crops are very profitable.

In the past, these crops were grown chiefly as cleaning crops in the rotation. Cultivation between the rows killed the weeds which had been encouraged by previous corn crops. The use of selective weedkillers in cereals has changed things, and today cash roots are important as break crops, giving the land a rest following year after year of cereal growing.

The feeding roots which were once grown in very large quantities for livestock have been replaced, partly by the forage crops which are cheaper and easier to grow and partly by beet tops which are a useful by-product of a profitable cash crop. New methods of drilling have made it possible to grow swedes and turnips more cheaply. At the moment there is more interest in these crops in the north and west of Britain.

The more common forage and root crops include the following:

Sugar beet, mangels, fodder beet	(beet family)
Turnips, swedes, kale, cabbage, rape, mustard	(brassica family)
Potatoes	(solanaceae)
Carrots and parsnips	(umbelliferae)

Chapter 6

Root Crops

IN RECENT years, root growing has been changed by several new developments:

Mechanisation is altering the growing, harvesting, and later handling of roots. Potato planters, precision root drills and mechanical thinners make it possible to cut down the labour needed for these crops. Complete harvesters for both roots and potatoes have the same effect at the other end of the growing season.

Prepacking and the retail sale of potatoes and other roots for human use make quality very important today. Good growing, careful handling, and quality control need to be watched all the time, and the grower can no longer sell his produce without taking these things into account. There are controlled grading standards.

Weedkillers have been developed for most of the root crops. They can be applied before sowing the seed, or before the plants come through the ground (pre-emergence) or after the plants appear (post-emergence). If used with care, they are effective and help to cut out work and trouble in the crop later.

Yields of the root crops are given in tonnes; in the case of the cash crops what matters is the weight of *saleable* roots. Storage in buildings is often necessary, and increasingly today these are expensive, specially designed buildings. Storage in clamps in the field is much less common. For information on calculating storage space for crop produce, see page 138.

It is not difficult to recognise the different root crops when they are growing, although there are several types and many varieties. The diagram opposite gives some idea of the way these crops grow in the soil.

POTATOES

This is the main cash root crop grown on our farms. It is an expensive crop to grow, needing expensive seed, good soil, heavy manuring, spraying, special cultivations and complicated harvesting and handling. The Potato Marketing Board regulates marketing and production, but the

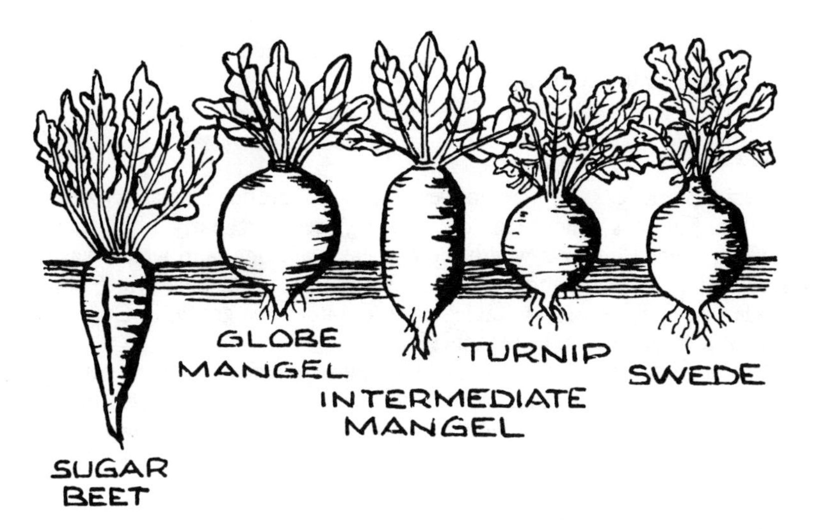

Fig. 23. The shapes of common root crops.

farmer has to sell his crop on the open market at the best price he can get—and this varies considerably, even in one season.

Potatoes can be grown on any soil which will work deeply, and which is not too heavy or stony. Potatoes (with oats and rye) is one of the few crops which can be grown on acid land short of lime, although it grows better on land which is not too acid. This crop does well on land with plenty of organic matter; it is an important crop in the Fens.

Potatoes suffer from a number of pests and diseases; it is particularly important not to grow them too often on the same land. A large part of the crop is now sold through the shops in pre-packs and for this reason sorting, grading and very careful handling are becoming more important.

The main types of potato growing are:

Seed Potatoes

To avoid infection with the virus diseases which are carried by aphis (greenfly), seed potatoes—tubers for planting—are grown in special districts. These are well away from main crop potatoes, and usually on high land where the aphis is not found. Seed potato crops are very carefully inspected for signs of disease, and are sold under certificate in various grades. Seed potatoes are grown in Scotland, Ireland, Wales and some western and northern parts of England; from these districts they are sold all over England. An ideal seed potato is about the size of a hen's egg.

Growers sometimes keep some of their own potatoes for seed, but because of the danger of spreading virus diseases this must not be done for more than one or two years, according to the district. It is then known as 'once grown' or 'twice grown' seed.

Early Potatoes

Special varieties are planted very early to produce small crops of immature potatoes which fetch high prices. These 'first early' potatoes are grown in special districts where there is easy working land which warms up early in the year, and protection from frosts and cold wind; some seaside districts are very suitable. Early potatoes are grown in Cornwall, South Wales, Kent and the Channel Islands.

Second early potatoes are planted later and are ready for lifting between the first earlies and the main crops. Their yield is midway between the two and the crop is sold at a time when prices are going down towards the normal prices paid for late potatoes.

Maincrop Potatoes

The crop is found throughout this country, but more are grown in certain arable districts where farmers specialise in potato growing (see the map on page 49).

POTATO GROWING

Types

Apart from the separate groups of varieties—earlies, second earlies and main crops—there are red potatoes and white potatoes. Some varieties are better for processing, for chipping, and for making into crisps.

Seedbed

This must be clean, level, and deeply worked—to give plenty of fine tilth for all the work which is to follow. In dry districts it is important to lose as little moisture as possible from the seedbed.

Fertilisers

Potatoes do best with heavy manuring and plenty of organic matter in the soil. Dung is usually given to this crop if there is any available, and should be ploughed in during the autumn; the average rate is 40 tonnes per hectare (15 tonnes per acre).

There is no need for lime—potatoes will grow on sour land.

(Refer to page 65 for information about normal rates of fertilisers.)

All the fertilisers are applied to the seedbed before planting, and may be distributed either before or after the ridges are opened for planting.

Chitting

Seed potatoes may be planted chitted (sprouted) or unchitted. Chitting is always done for earlies and sometimes for maincrops. For this, the seed has to be laid out in shallow boxes (trays), kept at the right temperature (2°–6°C: 36°–42°F), and exposed to light in buildings—all of which is expensive. Using chitted seed gives higher yields and quicker growth. It needs special handling when it is planted, so that the sprouts are not knocked off.

Planting

This job is done either by hand or by machine. The land is ridged up, the seed is spaced along the bottoms of the ridges and the ridges are split back to cover the seed. Normally the rows are about 0·75 m apart, with the seed spaced 30 cm apart in the rows, using about 2½ tonnes per hectare (1 tonne per acre). Early potatoes are planted closer, using rather more seed–up to 50 per cent more.

When seed is planted and covered, the ridges can be rolled down.

Early potatoes are planted as soon as possible in the spring, from January to March according to the district.

Maincrop potatoes are planted March/April, but not later than mid-April.

Weed Control

As the crop grows, it is possible to cultivate between the rows, to stir the soil. This helps to control weeds but spraying with selective weedkillers is the main method of weed control now.

Earthing-up

After cultivation and hoeing, the crop should be earthed-up finally before the tops meet across the rows. This ridging must be done carefully, to cover the tubers as they grow and to protect them from blight and greening.

Disease Control

Potato blight is the commonest disease of the crop, and is always worse in wet, warm seasons; it is to be expected from July onwards. When

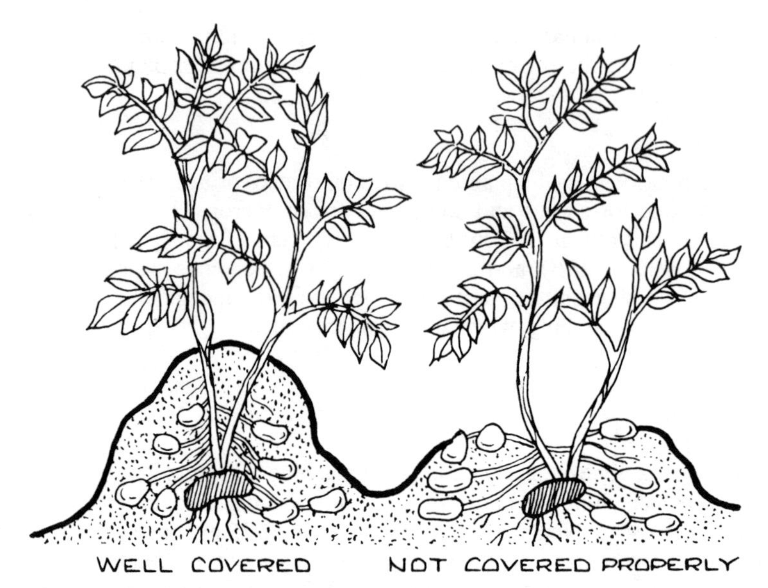

Fig. 24. Ridging up potatoes for protection against blight and daylight.

trouble can be foreseen, spray (or dust) the crop with chemicals to prevent blight on the leaves. The crop may need to be sprayed several times, according to conditions in the district. If it is infected with blight towards the end of the growing season, the haulm (potato tops) should be cut or burnt off using chemicals. This prevents the blight from the tops infecting the tubers in the ridges below—which will make them discolour and later rot.

Harvesting

Potatoes are ready to lift when the tops have stopped growing and died down, or have been cut or sprayed off. The skin on the potatoes is set, and does not rub off easily.

Potatoes are lifted with a digger or harvester. Hand picking is needed after the digger, and the crop is then carted away. The complete harvester does the whole job of getting the potatoes from the ground and away. It is important to treat potatoes gently, and to avoid damage in lifting and handling.

Storage

Potatoes are commonly stored in special buildings. Potatoes going into storage should be dry, sound and healthy, and free from damage. If possible, they should be allowed a few days to warm up and 'cure'. In the store, potatoes are protected against frost and wet. Later, they are taken from store and graded (riddled) for sale.

Potatoes are sorted as follows:

Chats (small) For feeding to livestock.
Ware Sound clean potatoes of good shape, for sale.
Oversize (large) Sold separately.

Yield

	Average	*Good Yield*	*Our Farm*
Tonnes of saleable tubers per acre	13	17	
Tonnes of saleable tubers per hectare	32	42	

Uses

Most potatoes grown in this country are sold for human consumption. More and more are being sold packaged, in transparent containers, and this is the chief reason why potatoes must be sold off the farm in good undamaged condition.

Some are used for processing (crisps, drying) and other industrial uses are developing. Surplus potatoes are sold back to farmers for stock feeding.

SUGAR BEET

Sugar beet is one of the crops grown *on contract* by the farmer. The contract is made between the farmer and a firm named British Sugar. This means that a stated quantity of beet will be accepted by British Sugar and paid for at a standard price, if it is of satisfactory quality.

Beet is a profitable crop if properly grown. It takes a good deal of care to grow it well—thorough cultivation, clean land, the right fertilisers, a sound knowledge of crop growing—and it must be harvested and handled properly. It is a crop which needs deep clean soil, well supplied with lime, with a level surface and which can be cultivated easily. Land which is wet, rocky, or too heavy is not suitable—the light and medium loams are best.

Where beet can be grown depends on the position of the sugar beet factories. These are not found in every county, but are concentrated in the

East. To avoid trouble with pests (mostly the sugar-beet eelworm), beet must not be grown too often on the same land.

The roots are topped before lifting, and these tops are a useful food for livestock. So also is the sugar-beet pulp which is returned from the factory. As well as being a good cash crop, sugar beet produces something of use for livestock.

Seedbed

This must be flat, level, fine, firm and clean. It must be deeply cultivated and not too dry, so that the seed can germinate evenly and grow quickly.

Deep autumn ploughing with a one-way plough is a good beginning, followed by the least possible cultivation in the spring.

Fertilisers

To produce a heavy crop, beet must be fed well. Dung is often given to this crop, and should be ploughed in during the autumn; if too fresh in the seedbed, it will make the roots grow fangy (forked).

Lime is important to this crop—beet will not grow on sour land.

(Refer to page 65 for information about normal dressings of fertilisers.)

Agricultural salt is commonly used, or kainit, to supply sodium which this crop needs. It should be applied well before the seed is sown.

Seed

Most sugar beet seed now is *monogerm*, containing one seed only in each piece. *Multigerm* seed, with more than one seed in each piece, is little used now. Beet seed is normally pelleted, surrounded by a hard chemical substance so that all the pieces are the same size; this helps to space the seed accurately in the soil.

Sowing

Spacing the seed at regular intervals along the rows needs precision, and most beet is now drilled with a precision drill. It is common now to 'drill to a stand', with the seed spaced at 12 cm or more in the row, so that the plants will grow as evenly spaced as possible, without hand-work. If there is plenty of hand labour at the right time, the seed can be sown more thickly, at about 7·5 cm, and the plants thinned out later.

It is drilled in wide rows, 2–2·5 cm deep, at about 10–16 kg per hectare (4–6·5 kg per acre) of pelleted seed in March/April.

Spacing

We aim to get about 74,000 plants per hectare (30,000 plants per acre) to get the best possible yield of beet.

The British Sugar Corporation provide tables which show the plant spacing in the row, at various row widths, to get as near to this as possible.

Hoeing

The crop should be hoed between the rows as early as possible, and hoed again as needed while the beet grow, using tractor hoes. Hoeing is mainly for weed control, and continues until the leaves are meeting across the rows. There may also be some hand hoeing in the rows to remove weeds, if there is enough labour.

Spraying

Sugar beet is normally sprayed for weed control. Chemicals may be applied in three ways:

Pre-sowing—worked into the soil before drilling, mainly to deal with grass weeds.

Pre-emergence—applied about the time of drilling or soon after. This may be band spraying (just down the rows) or overall spraying (over the whole soil surface).

Post-emergence—after the plants have come through the soil. This spray gives a chance to clear up weeds if the earlier sprays have not worked well. It may be either band sprayed or overall.

It is common also to spray beet during the summer against the aphis (greenfly) which spreads the virus yellows disease.

Bolters

Sugar-beet plants which are checked in their growth—by cold weather or bad growing conditions—produce a large top growth and try to flower and set seed in their first year. Some varieties of beet do this more often than others. The roots of bolters are tough and contain little sugar, and should not go to the factory.

A major problem now is 'weed beet'. These bolt and produce seed in crops of beet. There is a danger of this seed growing as a serious weed of other crops—as well as spreading disease to sugar-beet crops.

Harvesting

Lifting starts in September, but beet continue growing into October or November and are harvested until some time in December. Deliveries are controlled by the factory, which issues permits to the growers to make sure of a steady supply of beet. Most sugar beet are lifted with a complete harvester which tops, lifts and cleans.

It is important to make sure that beet are topped in the right place (too low or too high both mean a loss to the grower), that they are as clean as possible, and that any lifted beet are protected from frost.

From 1983 the growers' contract with British Sugar is based on a *quota* of so many tonnes of beet, the price varying according to quality, sugar content, etc., as set out in a contract which is arranged each year. To understand the system of payment, it is necessary to study the details of the current sugar beet contract.

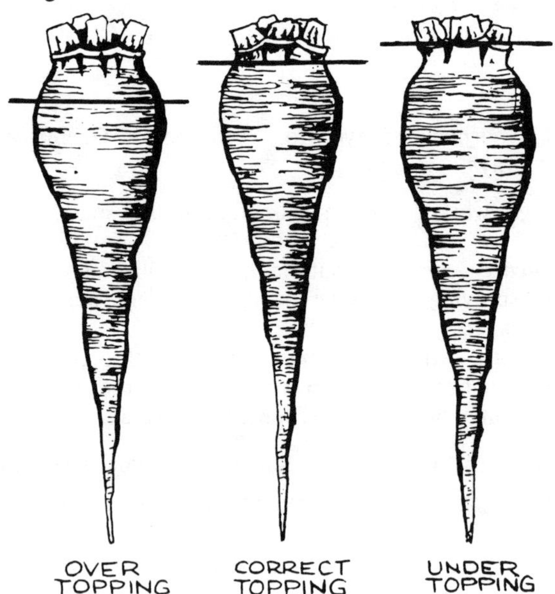

Fig. 25. Topping sugar beet.

Yields

	Average	*Good Yield*	*Our Farm*
Tonnes of roots per acre	14	18	
Tonnes of roots per hectare	36	45	

(Sugar-beet tops weigh about the same as the roots.)

Uses

Nearly all sugar beet grown in this country is factory processed to produce sugar. Sugar beet can be used for feeding, in just the same way as some types of fodder beet (see page 107) but this is not usual.

Beet tops are either fed on the field where they are grown or carted off and fed in grass fields or in buildings. They are a good food for cattle or sheep, but to avoid poisoning they must be allowed to wilt for a week or two before being fed. Beet tops can be made into silage.

Sugar-beet pulp comes from the factory in various forms. Wet it is a bulky food to replace roots; dry it is an energy food to replace cereals.

MANGELS

This crop—also known as mangolds or cattle beet—is commonly grown for feeding to livestock on the farm. With good conditions it is possible to grow a very heavy yield, but like all root crops it takes more labour than can sometimes be spared today.

Mangels do best with warm and sunny conditions, and for this reason are more common in the South and East than in the wetter parts of the country. A medium loam suits them best, but they can be grown on heavy land.

Less mangels are grown now than in the past, but it is still common to find a small area on many farms where livestock are kept.

Types

Most British varieties are low in dry matter, but there are some of medium dry matter.

Colour varies, and may be red, orange, yellow or white. Shape may be either globe, which is round, or intermediate, which is longer.

Seedbed

As for other root crops, a fine seedbed is needed, smooth, firm, flat and clean, with moisture below.

Fertilisers

Lime is important to mangels, just the same as sugar beet. Mangels will not grow on sour land.

(Refer to page 65 for information about normal dressings for fertilisers.)

It is common to give a dressing of dung for this crop. Agricultural salt can be used and should be applied well before the crop is sown.

Sowing

The seed is just the same as sugar beet seed, and should be drilled in the same way—normally with a precision drill. It is drilled in wide rows, 2–2·5 cm deep, at about 7·5–12 kg per hectare (3–5 kg per acre) in April/May.

Hoeing

Mangels are treated much the same as sugar beet, and hoed for weed control between the rows. Hand hoeing may be done if needed.

Spraying

The use of chemicals for weed control and for pests such as aphis is the same as for sugar beet.

Harvesting

The roots are ready to lift in October, and the job should be completed before there is too much hard frost. Mangels are lifted with an ordinary root lifter or harvester, but they are easily damaged.

Storage

Mangels should not be fed as soon as lifted, and are usually stored until the end of the year or into spring. They can be stored in buildings, or in field clamps, covered with plenty of straw.

It is common now to clamp with the tops on, but the tops may be cut off.

Yields

	Average	*Good Yield*	*Our Farm*
Tonnes of roots per acre	24	30	
Tonnes of roots per hectare	60	75	

Uses

Mangels are commonly fed to dairy cows, to other cattle and to sheep. Pigs will eat a few, but mangels are too watery and bulky to feed in any quantity. Feeding fresh mangels can cause trouble in any stock—leave them until the New Year to ripen.

FODDER BEET

This root crop comes between sugar beet and mangels—in shape, use, and feeding value. The many varieties of this crop range from the high dry-matter type which sits right down in the soil (like sugar beet) to larger and more watery roots which grow well out of the ground (like mangels); with intermediate types that combine good yield and reasonable feeding value.

Fodder beet is manured, cultivated and harvested just like mangels or sugar beet. It is used mainly for feeding to cattle but is not grown much today. Previously, it was used widely for pig feeding, at a time when other feeding stuffs were in short supply.

TURNIPS AND SWEDES

These two root crops are very much the same in their uses and the way in which they are grown. Swedes have a higher proportion of dry matter, and thus a higher feeding value, and will store for feeding during the winter period.

Turnips have hairy leaves, which grow direct out of the top of the root; colour may be white or yellow. Swedes have smooth leaves growing on top of a 'neck'; flesh may be white or yellow and skin green, bronze or purple.

Both of these roots do better under damper and cooler conditions than mangels, and for this reason are grown more in the North and West. They will do well on light to medium soils. In the past these crops were often grown on ridges, but this is less common today.

Both turnips and swedes are grown less now than in the past. The use of precision drills for sowing these crops has made it easier and cheaper to grow useful yields—and thus in some districts these crops are widely used for livestock.

Types

Turnips	White flesh	Earlier, quick growing, do not keep so well
	Yellow flesh	Later and hardier; keep better
Swedes	Purple skin	Earlier, less hardy
	Bronze skin	Intermediate type
	Green skin	Later and hardier, more solid flesh

Seedbed

Same as for other root crops, but particularly fine (because the seed is small) and with moisture below. The seed is drilled at 3·7–5 kg per hectare (1·5–2 kg per acre) or less for graded seed.

Fertilisers

Lime is important, and these crops will not do well on acid soil, which leads to trouble with Finger and Toe disease.

(Refer to page 65 for information about normal rates of fertilisers.)

Sowing

The seed is small and each one produces only one plant. It is drilled in rows in May/June. Quick-growing turnips can be sown later, until well into August.

This is a good crop to sow with a precision drill, as the seed is a regular size. This method is now usual as it allows hard work to be reduced or cut out altogether, which makes this a much cheaper crop to grow.

The crop is hoed between the rows until the leaves meet during the summer. Chemicals are used for weed control.

Harvesting

Turnips (and sometimes swedes) are commonly eaten in the ground by sheep. It is more usual to lift swedes in the autumn and store them. They can be stored in clamps or buildings, well protected with straw against frost.

Yields

	Average	*Good Yield*	*Our Farm*
Tonnes of roots per acre	18	25	
Tonnes of roots per hectare	45	60	

Uses

Turnips and swedes are fed to cows, other cattle, and sheep. They can also be sold for human consumption.

CARROTS

Formerly grown as either a market garden crop for human consumption, or on a field scale for feeding to livestock, carrots are now grown on a large scale as a cash root crop—for sale to processors, to the vegetable trade, or sometimes as a direct sale to the public. Surplus carrots can still be used for animal feeding.

It is a crop which needs good light soil in order to produce clean, good quality roots.

Types

Carrots vary according to shape and length, and which is grown depends on the market. The shorter stump-rooted types are better for canning; the larger, longer types for sale in shops, for slicing and dicing. The very large and long types are not found today, as they need very deep soil.

Seedbed

The same conditions are needed as for sugar beet and some other roots—flat, level, fine, firm and clean. It is important that the carrots grow steadily, without check. Avoid compaction of the soil, and this means as few cultivations as possible, to allow the carrot root to penetrate and grow to its best size and shape. Poor shape means poor quality.

Fertilisers

Lime is important—carrots will not grow properly on acid land. Refer to page 65 for information about normal rates of fertilisers, which will vary according to soil analysis and previous cropping. This is a crop which may need some of the trace elements, particularly on the light, sandy soils.

Sowing

Seed is usually precision drilled, and often carrots are grown on a 'bed' system—sometimes 9 rows in a bed, separated by 0·5 m wheelings—which varies in size.

Spraying

For weed control, chemicals are used both pre- and post-emergence. Chemicals are also used against pests, including carrot fly, aphis and cutworm in the soil.

Rotation

This is a crop which should not be grown too often on the same land—not more than one year in five, otherwise there may be disease problems.

Storage

Carrots can be stored in buildings, but it is common to store them in the soil, either ridged up or strawed over. Before sale, carrots are cleaned and it is common now to wash them.

Yields

	Average	*Good Yield*	*Our Farm*
Tonnes of saleable roots per acre	12	18	
Tonnes of saleable roots per hectare	30	45	

OTHER ROOT CROPS

Red beet (beetroot), radishes and parsnips are grown as field vegetable crops. Any surplus red beet or parsnips, not needed for sale, can be fed to livestock; cattle will make good use of them, and parsnips are a useful food for pigs.

One form of chicory is used sometimes as a herb in grass seed mixtures, when the leaves are grazed. Another type is grown as a salad vegetable. A small acreage of chicory is grown in Suffolk and Cambridgeshire for the roots, which are used to mix with coffee. This crop is grown and harvested just like sugar beet.

Artichokes are grown in small quantities for pig feeding; they produce a heavy crop of tubers and also a good deal of top growth. This crop is usually grown in odd pieces of land and provides cover for game birds. After the first planting it is left in the same land year after year, and needs very little attention.

THINGS TO DO

1. Measure the row widths, spacing in the rows, and find out the seed rates of the main root crop grown on your farm or in your district.

2. Measure the weight of individual roots, and find out the crop yields of the main root crop grown on your farm or in your district.
3. Study the details of a sugar-beet contract or the regulations of the Potato Marketing Board; whichever is the main crop grown in your district.
4. Study the harvesting methods used for root crops in your area.
5. Visit a sugar-beet factory, or see grading and processing methods used for potatoes or other root crops.
6. Study the information from British Sugar about the feeding uses of sugar-beet pulp in its various forms.

QUESTIONS

1. Which are the main root crops grown in your district, and why are they grown?
2. What quantity of seed is used for potatoes, and what are the main grades of seed?
3. Where are the sugar-beet factories in Great Britain—in which counties?
4. What different types of seed are available, and how does the price compare, of mangels and swedes?
5. What are average yields and prices of the main cash root crops?
6. What is the current price of sugar-beet pulp, and how does its price and value compare with a cereal such as barley?

Chapter 7

Forage Crops

OTHER THAN grass, there are a number of crops which provide green food for stock—either for grazing in the field, cutting and feeding elsewhere, or for silage. These crops come from a number of plant families, and are used at different times and in various ways. Generally, they are cheaper to grow than roots, take less labour, and have a higher feeding value. The forage crops include the following:

Brassica crops	Kale, cabbage, rape, turnips, mustard, kohl rabi.
Leguminous crops	Vetches, lupins.
Cereal crops	Rye, maize (and occasional spring grazing of cereal crops).

The first use of forage crops was for sheep feeding, when 'hurdled flocks' of lowland sheep were folded over crops of rape, kale, cabbage and kohl rabi grown specially for them. Such use of these crops is rare today—it took too much labour and used land which now grows cash crops—although some lambs are fattened in autumn and winter on crops of rape or other green crops.

A much more important use of the forage crops today is for the feeding of cattle, particularly dairy cows. The green food produced from these crops is used to feed cattle in the autumn and winter, when grass has stopped growing—so they can go on grazing well into the winter. Kale is the most popular crop for this purpose. Farmers in some districts still like cabbage—which takes more labour but produces a very heavy crop.

The ideal would be one or more crops which would give enough green food for grazing (on land light and dry enough) or for cutting and carting to stock in yards, right through the winter until the grass came again. This list shows how some of these crops are used:

Maize	September
Rape	October
Marrow stem kale	November/December
Thousand head kale	January/February
Rye	March
Italian ryegrass	March/April

The gaps—when no growing green crops are available—can in practice be filled by feeding silage (which can be made from green crops, grasses and clovers) or by the use of roots, hay and straw.

To ration livestock, and to calculate how long supplies of food will last, it is often necessary to calculate yields per area of forage crops (and also grasses and clovers).

A rough and simple method for the crops which cover the ground fairly closely—such as rye, ryegrass, or any other grasses and clovers—is to cut and weigh the quantity on several patches of one square metre each (in different parts of the field). ½ kilo per square metre is roughly equal to 5 tonnes per hectare (2 tonnes per acre) and from an average figure in the field a quick calculation can be done.

KALE

In many ways, kale is the most important and the most widely grown of the forage crops. It is cheap and easy to grow, and if grown on good land or fed well it will produce a tremendous weight of stock food. The other advantage of kale is that it can be sown at different times, and various kinds used, so that it provides keep for a long time in the autumn and winter.

The most valuable part of the kale plant for feeding is the leaf, which contains more protein and less fibre than the stem.

Types

Marrow Stem is the most popular and gives the heaviest yields. It produces a thick long stem with leaves above. The whole plant can be eaten; it is either grazed in the field or cut and carted to stock. This kale is not hardy and in many districts should be used before the end of the year.

Thousand Head gives lighter yields, but produces a larger proportion of leaf and a thin, short, fibrous stem. This kale is hardier than marrow stem and is used in the colder weather well into the following year.

Sowing

Kale needs a fine, firm seedbed like the root crops. It can be drilled (or broadcast) at any time from March to June, at 5 kg per hectare (2 kg per acre). Seed is drilled 2–2·5 cm deep, in rows about 0·5 m apart. Seed is dressed to protect the seedlings against flea beetle.

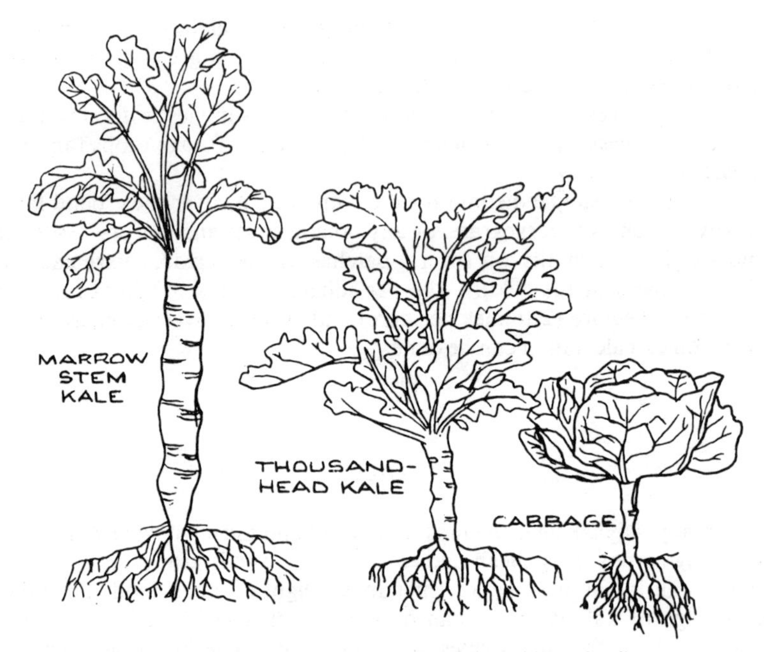

Fig. 26. The main forage crops of the cabbage family.

Fertilisers

The land must not be short of lime. This crop makes good use of a dressing of dung and also does well on ploughed up grassland. Nitrogen is important to give a heavy yield of kale.

(Refer to page 65 for information about normal rates of fertilisers.)

Kale can be left unthinned in the rows—the simplest way to grow it—or it can be chopped out, which gives fewer and larger plants. The crop is hoed between the rows, and may be sprayed to control weeds; when the plants meet in the row they smother rubbish.

All types of kale can be grazed where they are grown. The crop is also harvested—either by hand, or using a mower or a forage harvester.

Sometimes, the odd crop of kale can be sold as greens for human consumption in a season when other green crops are very scarce.

Yields

Marrow stem kale will easily produce 30–50 tonnes per hectare (12–20 tonnes per acre), 50 per cent more under good conditions. The other types give less.

CABBAGE

Cabbage for feeding to livestock—sometimes called 'cattle cabbage'—is grown in some districts. There are several varieties, some the same as garden cabbages and others, like the Drumhead, grown specially for stock feeding.

The head of a cabbage forms a firm, tightly-packed ball of leaves on a small fibrous stem. Its feeding value is very similar to that of kale leaves.

This crop is grown in the same way as kale and needs the same seedbed and general conditions. Seed is sown in rows 0·5–0·7 m apart, 1–2 cm deep, at 3–5 kg per hectare (1–2 kg per acre). Sowings may be made from late March until May or even later, and, like kale, the seed should be protected against flea beetle. Plants are chopped out to about 0·25 m apart.

Another method of growing this crop, much less common, is to sow seed in a seedbed, and plant out the small cabbages when they are strong and sturdy.

Cabbage is harvested by hand-cutting, and is then carted away to be fed to stock. It is left in the field until needed, as it does not store well.

FORAGE RAPE

This crop is like turnip or swede tops without a thick root below. It is quick growing and of good food value, being rich in protein. Rape is not found much in southern and eastern districts, but is more popular in the West and North, where it is commonly grown for sheep feeding.

Rape is grown on its own, either in rows or broadcast, or often mixed with other forage crops—rape and ryegrass; rape, turnips and kale; etc. It can be undersown in a cereal crop for autumn grazing, or sown on the stubble as soon as the corn is cleared.

Seed is sown from May onwards until August, at 5–12 kg per hectare (2–5 kg per acre), drilled or broadcast.

RYE

As well as being a cereal crop for use on light soil, rye is sometimes grown for grazing in the spring, particularly for dairy cows. It is the earliest of all the forage crops, producing a useful high quality grazing, and if rested well give more grazing later—a total of two or three times in the season. To get this very early growth, rye must have a dressing of nitrogen at the beginning of the year.

Ordinary grain varieties are not suitable for this use, and there are special grazing varieties.

For this early spring use, rye is drilled in the same way as other winter corn, at 150 kg per hectare (60 kg per acre), at any time from August onwards—the earlier the better.

MAIZE

There has been more interest in this crop for silage-making during the last few years. New varieties have been developed which suit our conditions and forage harvesters can deal with this crop quite well.

There are some snags in growing maize. Rooks are very troublesome, digging up the seeds and pulling up the young plants, and harvesting can be difficult in a wet and a late season. Maize will not stand the frost and is not grown in the colder parts of the country where the growing season is too short.

Sowing

Seed is drilled at 34–45 kg per hectare (14–18 kg per acre), 2·5 cm deep in April/May.

Harvesting

As a green food, maize can be cut and carted when it is well grown, from August onwards. It is a useful crop in a dry season. For silage, it is best to wait until the seeds on the 'cobs' have developed and reached the 'cheesy' stage in September/October.

MUSTARD

There are two separate types of this crop which are grown for different purposes.

White Mustard: is a quick growing plant used either for feeding to sheep or for ploughing into the land (as green manure). It is sown in rows or broadcast at 15–22 kilos per hectare (6–9 kilos per acre) at any time from April to August.

Black Mustard: is grown in the Eastern Counties on contract, to produce seed for grinding to make mustard for the table.

OTHER FORAGE CROPS

A large range of forage crops were grown in the past, for such special purposes as sheep feeding. Some of these crops are still available, although not common now, and there are other forage crops which have been developed and which may come into more general use.

Stubble Turnips: are quick growing turnips, producing much leafy top and not so much root, which can be sown in late Summer and into cultivated stubbles after cereals are harvested. The crop produces plenty of useful feed for sheep, which is eaten in the field.

Fodder Radish: is a fairly new, very quick-growing crop, which produces a great weight of top. It is grown in some districts for feeding to sheep and cattle, in the field.

Kohl Rabi: is sometimes grown for sheep and may be seen on farms where rams are bred. It was once widely grown for folded flocks of sheep. It is like a large turnip root growing above the ground, green in colour, with a bunch of leaves as the top. It is also grown as a garden vegetable.

Vetches: are also known as tares; this plant is one of the legume family. It is seldom grown now, but was used at one time in mixtures such as oats and tares. It produced a great bulk of green material, with long trailing stems up to 2 m long.

THINGS TO DO

1. See NIAB leaflets and seed catalogues from local and national seedsmen, to find out what varieties and types are on sale.
2. See different forage crops growing in the field.
3. Estimate the yield in the field of these crops.
4. See samples of silage made from forage crops, and inspect it for quality.
5. See samples of seed of as many forage crops as you can.

QUESTIONS

1. Which plant family provides most of the forage crops other than grass?
2. Which is the most valuable part of the kale plant for feeding?
3. Which of the cereals are used as forage crops?
4. Which forage crops are used for silage making?
5. What methods are used to harvest the main forage crops grown in your area?

Chapter 8

Grassland

TWO-THIRDS of the farming land in this country is grassland—and more than half of the produce of British farms comes from grassland. Grass is our biggest crop.

A good part of this country is well suited to growing grass. It is the crop which produces more foodstuff per acre than any other. For those livestock which can use it properly—mainly cattle and sheep—it is the cheapest food that we can either produce or buy. It is the one crop which needs much less labour than any other. It is also the one crop which is most badly used.

HOW GRASSLAND IS USED

Grassland cannot be used directly for human feeding, as can many other crops. It is used to feed animals, which produce food for human use—milk, beef, or lamb—or other products, such as wool or leather.

Grass can be either fed direct in the field where it grows (grazed), or cut and carried to livestock which are kept elsewhere (zero grazing) or cut and preserved for use later (hay or silage).

TYPES OF GRASSLAND

Grassland can be any combination of these plants—grasses, clovers, herbs, weeds.

Grassland plants may be grown and managed in several different ways:

Catch Crops: Usually a single type of grass grown for a short time between two other crops, e.g. Italian ryegrass.

Short Ley: One or more grasses and clovers grown for one to three years.

Long Ley: One or more grasses and clovers grown for four to twelve years.

Permanent Grass: A mixture of grasses and clovers which has either been sown as a ley and grown for more than twelve years or which has grown naturally for a long time.

Most of our grassland is permanent grass—more than twice as much as leys. Some of this is on land which is wet, hilly, or difficult to cultivate and which must therefore be kept in grass rather than arable crops.

There are also many acres of very poor permanent grass—on hills, moors, and mountains—which is known as rough grazing.

GRASSES

There are hundreds of different grasses. Only half a dozen are commonly sown for farm use in this country, but there are several different varieties of each one.

The Main Grasses

Italian Ryegrass is a quick-growing, high-yielding grass. Seed is cheap, it is easy to grow, and it produces a crop of seed easily. It is used for both grazing and cutting, and is usually only grown for one- and two-year leys, seldom lasting more than three years. It starts to grow early in the spring, and is well liked by livestock.

Perennial Ryegrass needs good land and good feeding to do well. It is one of the commonest grasses, found in good permanent grassland and widely used in leys. It is particularly good for grazing, useful for cutting, and lasts a long time on good land.

It is a fairly early grass, producing plenty of growth in spring and early summer, but growing very little in July and August. It is hard-wearing and stays green in the winter.

Hybrid Ryegrass is a cross between Italian and Perennial.

Cocksfoot is a strong, deep-rooted grass which stands up to dry conditions. It can produce a great bulk of growth, but is not so digestible to livestock—nor so well liked—as the other grasses. It grows during the summer period when ryegrass is producing very little.

Timothy is not so strong as the last two grasses, and will not grow well in competition with them. It grows rather later in the spring, and also goes on growing during the summer period. It is a grass which is very well liked by livestock, and which stays green in the winter. It needs good land to grow well.

Meadow Fescue is commonly grown in mixtures with timothy, and is very like it in growth and uses. It starts to grow earlier in the year, also grows fairly well in the autumn, and stays green in winter.

Less Important Grasses

There are several other grasses, most of which are not commonly sown now. They are, however, often found in old permanent pasture and begin to grow naturally when grassland has been down for a few years.

These grasses fill up the spaces between the plants of the main grasses, and for this reason are sometimes called 'bottom grasses'. Some of them are well liked by livestock and add something to grazing, but none of them is so productive as any of the main grasses. Common grasses of this type are:

Crested Dogstail. A small grass found in old pastures.
Meadow Foxtail. One of the first grasses to grow in the spring.
Tall Fescue. Another early grass, which may be used more in future.
Red Fescue. A creeping grass, used in lawns.
Rough Stalked Meadow Grass }
Smooth Stalked Meadow Grass } Common and palatable grasses, liked by livestock.
Sweet Vernal Grass. Gives a very sweet smell to hay, but a bitter taste.

Weed Grasses

Agrostis (*Bent*): is the commonest grass of all in this country, found in nearly all old permanent pastures. Its stems run over the surface of the soil, forming a mat; this is its one advantage, as it makes a hard-wearing surface. It starts to grow late in the year, produces very little, and is not very much liked by livestock. In spite of the wear it gives, it is better replaced by stronger growing grasses.

Yorkshire Fog is soft, hairy, and of little use for livestock; it takes up space in grassland which would be better used by the main grasses.

Couch (twitch or speargrass) is an arable weed, with creeping stems, difficult to destroy. Onion couch is a different grass, troublesome in some areas.

Wild Oats and *Black Grass* are two of the worst arable weeds, which seed very freely.

Meadow Grass also has become an arable weed in some parts of the country.

RECOGNISING THE COMMON GRASSES

So many grasses are found growing commonly in this country that it is important to be able to recognise a few of the more useful. This can only

Identification of Grasses

Name of Grass	*How it Grows*	*Texture*	*Shoot*	*Auricles*	*Ligules*	*Leaf*	*Base of Stem*
Italian ryegrass	Tufted	Smooth	Round	Yes	Yes	Glossy below, dull, ribbed above	Red
Perennial ryegrass	Tufted	Smooth	Flat	Small	Yes	Same	Red
Cocksfoot	Tufted	Rough	Flat	No	Yes	Light green, rough edges	Light
Timothy	Tufted	Smooth	Round	No	Yes	Light green	Brown (sometimes swollen)
Meadow fescue	Tufted	Smooth	Round	Small	Yes	Same as Italian ryegrass	Red
Bent	Stolons	Smooth	Round	No	Yes	Long, narrow	Creeping
Couch	Rhizomes	Hairy	Round	Yes	Yes	Pointed, hairy	Sometimes red
Yorkshire fog	Tufted	Hairy	Round	No	Yes	Light, hairy	White, with pink veins

N.B.—Perennial ryegrass and meadow fescue have very small auricles which easily rub off.

be done by practice. There are two chief methods of telling one grass from another:

1. Flowering Heads

The heads, which are the flowers of a grass, appear during the summer. They can easily be seen in a crop left for hay, or on plants growing in hedges or other odd corners where they escape from cutting or grazing. It is also useful to be able to recognise them in hay. With a little practice, it is quite easy to know the heads of the more important grasses. The diagram below shows their general appearance.

To avoid confusing them, practise telling apart some grasses which are rather similar. For example: Perennial ryegrass, Italian ryegrass and couch are similar, but Italian has *awns* (long points) and couch has its flowers set crosswise on the stem, rather than end-on like the ryegrasses. Timothy, meadow foxtail and crested dogstail are similar, but foxtail appears very early in the year, and has a softer, silky, more tapering head then timothy, crested dogstail is much smaller, and is set on a shorter stem.

Make a Collection of grasses in the flowering stage. Get a bunch of each of the common grasses (including some of the weed grasses), dry them, label them, and use them for reference.

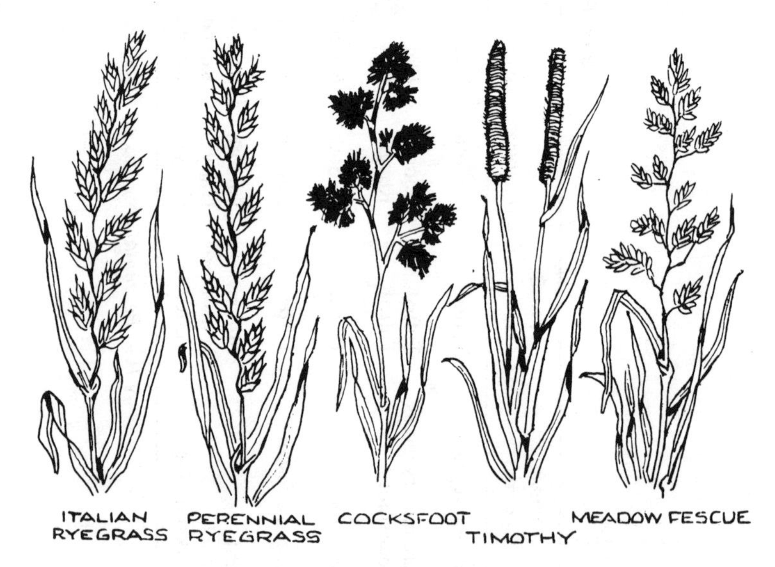

Fig. 27. Flowering heads of common grasses.

2. Leafy Parts of the Plant

Grasses can be recognised at any time when the plant is growing normally, so long as it is not too withered or damaged. There are a number of points to look for:

How It Grows: Most of our common grasses grow as single plants, each one forming a little tuft. Some spread from the original plant, by stems which creep either over the surface (stolons) or underground (rhizomes).

Rough, Smooth, or Hairy: Most of our common grasses have smooth leaves and stems, but a few feel rough to the touch (some even cut the hand) and some are hairy.

Shoot: may be either folded or rolled; it should be cut across and examined carefully.

Leaf: may be a particular shade of green, may have lines running down it, or may be of a distinctive shape.

Base of Stem: may be coloured (particularly in older plants) or may be swollen into a sort of bulb.

Auricles and *Ligules* are found in some grasses.

CLOVERS

Except where land is sour and short of lime, clovers are found growing naturally in permanent grassland. They are often sown in seed mixtures along with the grasses.

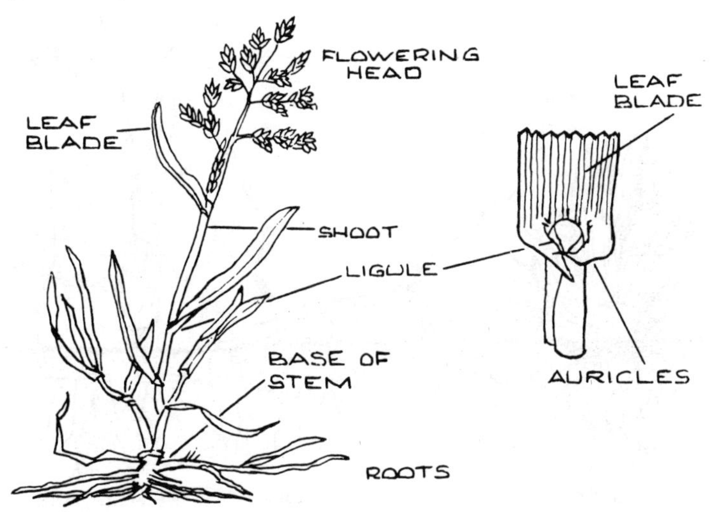

Fig. 28. Main parts of a grass plant.

All the clovers (and some other plants like them) are members of the legume family. They take nitrogen from the air, through the bacteria in the nodules on their roots, and much of their importance is this fact—they can feed themselves, and the grasses growing alongside them, with nitrogen at no cost to the farmer.

The use of clovers in grassland varies according to the type of grassland farming carried out. This is influenced by the price of nitrogen fertilisers. If you give heavy dressings of N fertilisers, this encourages the growth of grass and suppresses the clovers. If N fertilisers are very expensive, it may pay to use less and to encourage the growth of clovers which provide a supply of N to feed the grass. In recent years both systems have been in use.

THE MAIN CLOVERS

White Clover is used in many leys which are to stay down more than two years. It blends well with the common grasses and it creeps and covers the ground, forming a good bottom to a pasture. It is a useful plant for both cutting and grazing, and it is well liked by livestock. There are two main types:

(a) Creeping white clovers, such as the well-known wild white, have small leaves and creeping stems close to the ground. Their chief value is for grazing (particularly sheep).

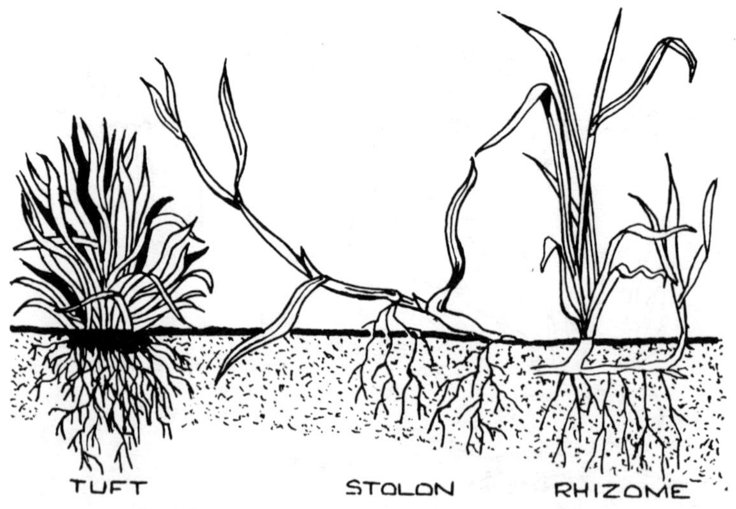

Fig. 29. How different grasses grow and spread in the soil.

(b) Tall-growing white clovers have larger leaves and longer stems and produce more food. These types are of more use for cattle grazing and for cutting.

Red Clover is used mostly in shorter leys, which are down for one or two years, but it is used less today than in the past. Red clover stands higher than white clover; the plants grow on their own and do not creep and cover the ground. There are two main types:

(a) Broad red clover does not really crop for more than one season. Usually sown with Italian ryegrass under a cereal, it produces a heavy growth the following year. It is often cut twice—in the summer and in the autumn, and is then ploughed in.

(b) Late-flowering red clover lasts from two to five years, according to the variety used. It flowers about two weeks later than the broad red, and is usually cut only once a year. It stands up to grazing better than broad red.

Lucerne is usually grown as a special crop on its own, or with a small amount of grass. It is very deep-rooted, stands up well to dry conditions and produces a great amount of green material; but not much is grown as it is not suited to grazing, and is best cut. It can be cut three or four times a year, and grows for a number of years.

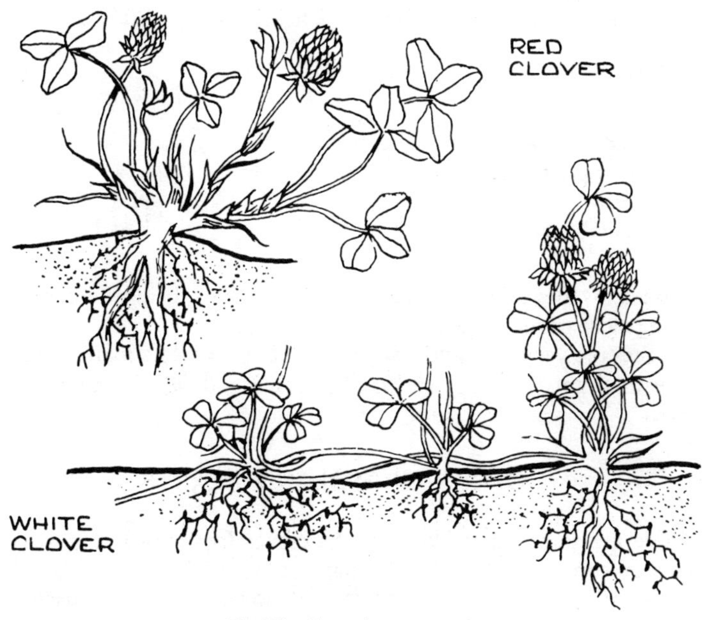

Fig. 30. How clovers grow.

It is important that the land is clean, well-drained, and not short of lime if this crop is to grow well. It is sown in April or May, and is either broadcast or drilled at the same width as a cereal crop. Lucerne will not grow properly unless the seed is dressed with a 'culture' of the bacteria which live in the roots and feed the plant with nitrogen. (Lucerne is not strictly speaking a real clover, but a related legume.)

RECOGNISING THE COMMON CLOVERS

There are a few clovers (and related plants) which you should be able to recognise. As with the grasses, it is possible to tell some of them apart by the flower, and by the leafy parts of the plant. The table below shows how they compare:

LEAFY PARTS	COLOUR OF FLOWER
Leaves with 3 leaflets to each	
Hairy plants	
Trefoil (pointed tip to the leaflets)	Yellow, small
Red Clover (red lines on the stipules)	Red, Pink
Hairless plants	
Lucerne (pointed tip and ragged edge to leaflets)	Violet, Blue (sometimes yellow)
White Clover (creeping plants, rooting along the stolons)	White
Leaves with 5 or more leaflets	
Slightly Hairy Plants	
Sainfoin	Pink, with red veins
Vetch (long tendrils at end of leaves)	Red, Purple

White clover is different from any of the others, being a creeping plant, growing close to the ground.

Vetches are very long and straggling, and grow into a matted tangle.

The diagram (Fig. 31) shows the leaves and other features of these plants.

OTHER GRASSLAND PLANTS

There are some other plants of the clover family (leguminosae) which have been commonly grown in the past, but which are seldom found now.

Alsike is rather like late-flowering red clover, and can be used on land which is wet and sour, where it may grow better than red clover.

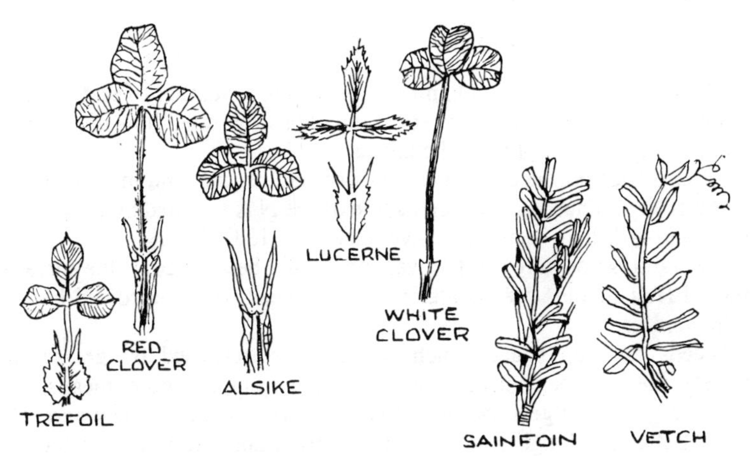

Fig. 31. Leaves of clovers and similar plants.

Trefoil is a quick-growing annual plant, with a yellow flower, which can be sown in a cereal crop to give stubble grazing.

Crimson clover is another annual plant, with a bright red flower, once used for sheep feeding.

Sainfoin grows rather like lucerne, but is only found on chalky soils in one or two districts.

Herbs are sometimes grown in grass seed mixtures, either scattered through a field, or in special 'herb strips'. Some people consider them a useful addition to the diet of grazing animals. The herbs used in this way include: chicory, plantain, burnet and sheeps parsley. Most of them grow naturally in old pastures.

PERMANENT GRASS

Any grass which has existed naturally for a long time, or a ley which has been sown down and not been ploughed for more than twelve years, is known as permanent grass. If used mainly for grazing, it is a permanent pasture.

Permanent grass usually wears better than a ley—it stands up to treading by livestock at times of the year when a ley might be 'poached' (trodden and spoiled) badly. If permanent grass is made up of good grasses and is fertilised and managed well it will produce just as well as leys. Unfortunately, much old pasture is poor, and the commonest grass found in it is Agrostis (see page 120). A pasture of this sort is late to start growth in the spring, and only produces a low yield of animal food.

IMPROVING OLD GRASSLAND

The quality of grassland can be improved in a number of ways. Liming will be necessary if the land is sour; the only way to tell how much lime is needed is by soil sampling and analysis.

Grass is not produced as a free gift of nature; the land must be fed. N, P and K are all needed, although some of the N is supplied by clover in the grassland—if there is enough clover (see page 133).

Wet land is not fully productive, and should be drained. If the surface is matted (full of old, dead material) it may need to be harrowed hard, or cultivated.

Weeds take up space which would be better used by grasses and clovers; they can be controlled by spraying, cutting and other means.

If old grassland gets too bad, it is not worth wasting money and effort on it. Better to plough it up, perhaps grow arable crops for a while, and sow it down to grass again.

LEYS

There are many different sorts of ley, sown for various purposes. The farmer selects a suitable seeds mixture and aims to produce a ley which will fit in with his system of farming and suit the needs of his livestock.

It is not only the seeds mixture which decides how a ley will develop. This is very much influenced by the type of land, and by the way in which the ley is managed.

Many different seeds mixture are used. Most of them consist of one or more grasses, sometimes with clover.

Short Leys are sown down for one or two years. A common practice once—not found so often today—was to use simple mixtures of Italian ryegrass and red clover for hay production. Sown down under a cereal crop (cover crop) it was cut for hay the following year, and then ploughed up to be followed by a crop of winter corn. Nowadays ryegrass is the usual grass for this sort of short ley, either for hay or for grazing. Seed mixtures for short leys are simple. For one-year hay production, a typical mixture would be:

5 kg ryegrass (Italian or perennial)
3 kg red clover

---

8 kg per acre (20 kg per hectare)

For one-year grazing, a typical mixture would be:

7 kg Italian ryegrass
7 kg hybrid ryegrass (cross between Italian and perennial)

---

14 kg per acre (35 kg per hectare)

Long Leys may be down for 3 or 4 years, or for longer periods of up to 10 or 12 years (after which they may become permanent grassland). Long grazing leys are usually based on ryegrasses, sometimes with a small amount of other grasses. A typical mixture would be:

5 kg late perennial ryegrass
4 kg medium perennial ryegrass
1 kg timothy

---

10 kg per acre (25 kg per hectare)

Mixtures including other grasses—cocksfoot, timothy and meadow fescue—are little used today, except on light, dry soils. White clover is only included in this sort of mixture if the use of N fertilisers is to be kept below 375 kg/ha (300 units per acre).

Commercial Seeds Mixtures

To get some idea of the mixtures which are commonly sold and used, see the seed catalogues of several firms. There is also a useful Ministry of Agriculture advisory leaflet on seed mixtures.

The End of a Ley

The useful life of a ley is finished:
When it has been down for the length of time intended.
When it can no longer produce what is needed from it.
When much of the good grasses and clovers have been replaced by inferior plants.

The usual practice is to plough up a ley (or permanent grass), so that the old turf is covered completely and rots down. It can then either be:

(a) Sown directly back to grass again—direct reseeding.
(b) Cropped for a number of years with arable crops (and perhaps sown down to grass again later).

Soil Fertility is increased by a ley which has been down for three years or more. A great amount of organic matter from the roots and the above-ground parts of the grasses is added to the soil. This is good for soil condition, and helps with arable cropping for a few years.

SOWING GRASS SEEDS

Grass must be treated like any other crop, and the land must be prepared properly. The chief needs are:

1. A fine, firm, clean seedbed.

2. Plenty of lime—the land must not be acid or clover will not grow properly.
3. Plenty of N, P and K (50 units of each) to give the seeds a good start.
4. Firm land, both before and after the seeds are sown.

The usual times for sowing grass seeds are March/April and July/August. The later sowing is often better in districts which have a dry spring.

Seed may be broadcast onto a rolled surface or drilled; drilling is better in dry districts. The best drill for the job is one with narrow row widths; if the usual cereal spacing is used, half the seed can be drilled one way, and the rest at right angles across it. After drilling, the land is harrowed lightly and rolled down firm.

Direct Seeding means sowing grass seeds on bare land, with the seedbed specially prepared for them. Sometimes a light seeding of cereal for grazing is sown with grass seeds to give some early keep for stock, before the grass is ready. This is known as a *nurse crop.*

The New Ley: When the seedlings first come through, stock can be put on the land to give it a good treading, which encourages the young plants. Light grazing of the first growth helps a new ley to become established, encourages the clovers, and makes the plants cover the ground, thus giving a good bottom to the pasture.

Weeds can be troublesome after sowing; they can usually be controlled by cutting over once or twice with a mower set not too close to the ground or by the use of sprays.

A new ley, as soon as it is growing properly, makes a very high quality food for livestock. It is particularly good for milk production.

Any grazing done in the first year should be short and sharp—livestock should not be kept on too long.

GRASSLAND MANAGEMENT

This is a complicated subject, and the management of any field of grass depends upon the type of grass, the type of livestock, the land, and the whole farming system.

There are a few simple rules, which are explained in the sections below, but grassland management, to be good, needs a great deal of thought, experience, and study.

How Grass Grows

Most grasses (and the clovers which grow with them) are perennial plants, and grow for a number of years. The growing points (where new

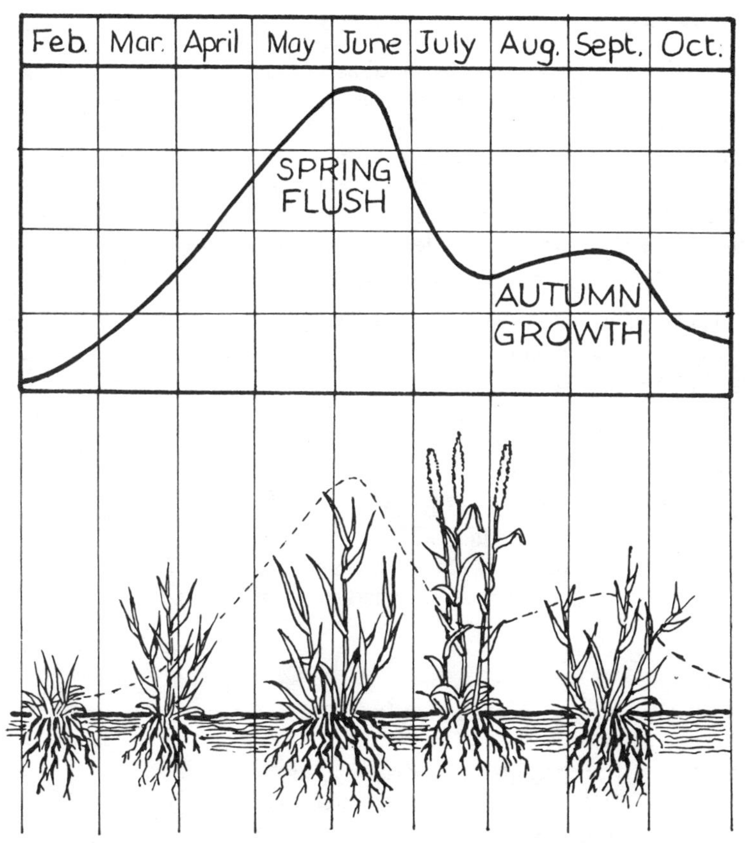

Fig. 32. How grass grows in the year.

shoots start) of these plants are at ground level, and so are not damaged by cutting or grazing.

In this country, grass does not grow the whole year round. It starts growth in the spring (earlier or later according to the plant and the variety), produces a high yield of good quality leafy material and then starts to put up stalk and flowering heads. If this is not stopped it sets seed, and the flowering heads, the stems and many of the leaves become dry and withered.

If grass is prevented from flowering and seeding, it produces more leafy material, of a higher quality, over the year. It produces most in the spring, rests for a time during the summer, and produces another lot of growth in the late summer/autumn period, as shown in fig. 32.

Feed Grass Well

Grass is a crop, and must be treated as a crop. Like other crops, it needs feeding in the form of lime and fertilisers.

Grazing returns some plant foods to the land in the animal dung and urine. If grass is cut and taken away (hay, silage, or zero grazing) more fertilisers must be used to make up for what is removed.

If grass is needed out of season—such as for very early grazing in the spring—it is necessary to apply fertilisers to encourage early growth.

(Refer to page 65 for information about normal rates of fertilisers.)

It is always best to give time for fertilisers to be washed into the land by rain before a field is grazed and this is particularly important with basic slag in powder form.

Let Grass Rest

Grass needs periods of rest, so that it can regain its strength, make more growth, and give most production.

The growth of the roots—through which the plants feed—depends on the amount of top growth. If grass is grazed down hard all the time (or kept short, like a lawn) the roots do not grow down to any depth, and there is very little production.

The right sort of management is simple enough with cutting. A grass field for hay or silage is rested until it is ready for cutting (see page 136) and rested again after cutting.

A field used for grazing needs resting in just the same way. After it has been grazed off properly, it should be rested, and then grazed again when it has made enough growth. The length of rest period varies with the time of the year. Growth is quick in the spring and early summer, slower in late summer, and speeds up again in the autumn for a short while.

It is often possible to rest a grass field in another way. It may be grazed one year, and cut the next. Or it may be grazed for a while in the spring, cut for silage during the summer, and perhaps lightly grazed again in the autumn. Changing the management in this way rests the grass, and helps to get the most production from it.

Prevent Seed and Stem

For grass to send up a stem, produce a flower, and set seed, takes considerable effort. Every time grass does this it is weakened and this reduces the production of leaf—which is the most valuable part of the plant.

It is important therefore to prevent grass used for grazing from making stem and setting seed. The longer this can be done, the longer the useful life of the grassland.

In dry seasons and on thin soils, grasses (and clovers) are more likely to produce stem and seed; fields which are *topped* (cut over) with some sort of mower remain fresh and give the best quality grazing.

For grazing by sheep—which need a very short pasture—gang mowers are sometimes used on good land. These cut close to the ground, and make a field like a lawn.

HOW LIVESTOCK GRAZE

All types of livestock graze differently. Their grazing affects the growth of grassland, and some types of grassland suit them better than others. It is a good practice to watch stock grazing—see what they eat, where they eat, and how they crop it off.

Sheep like short grass, and are not at all happy with anything longer. They bare their teeth and nibble right down to the ground. When there appears to be very little growth, sheep can still get a good feed.

Cattle prefer longer grass. They put their tongue round a clump of grass and crop it off with their teeth. Calves can eat shorter grass quite well, but not so short as sheep.

Pigs are very choosy in their grazing. They prefer clover, and some other plants such as chicory, and do not care much for grass.

Horses graze in small patches, and leave clumps of grass untouched, which after a time makes any field rough and ragged.

All animals avoid grazing where dung from their own kind has been dropped. This is one reason why it is best to rest grass, and if possible graze a field with different types of livestock.

If sheep follow cattle, they will eat the short grass which the cattle have left and which might otherwise be wasted.

To keep livestock as healthy as possible, and to avoid trouble with worms in the stomach and intestines, it is best not to keep any type of livestock too long on one piece of land, nor to graze them on it year after year.

THE MAKE-UP OF A PASTURE

Grass fields may consist mainly of grasses or a mixture of grasses and clovers. If you aim to use heavy dressings of N fertilisers, grass alone is best. If you aim to get some N from the clover, 60 per cent grass to 40 per cent clover is a good proportion. The more N fertiliser you use, the less clover there will be.

Management also will alter this balance. If a field is grazed early in the spring, the clover will increase. If it is left for a while and grazed later in the spring, the grass will be encouraged. If a strong growing grass such as cocksfoot is under-grazed at any time and allowed to get away, it will after a while swamp out the other grasses.

If a field is grazed with the same number of stock all through the season—over-grazed at some times and under-grazed at others—it will become rubbishy and full of weeds.

SYSTEMS OF GRAZING

There are many ways in which pastures can be used for livestock.

Uncontrolled: Stock allowed to roam about over the land, and eat what and where they like. This is bad for grassland, and does not make full use of it.

Set Stocking (or Field Grazing): A certain number of stock put into a grass field, and left there until it is eaten off. They are then removed and the field rested.

Paddock Grazing: A field is divided into small paddocks. All the stock are put into these in turn, so that each paddock is grazed off quickly (in two to three days) and then rested. This produces more from a field than the last two methods.

Folding: The land is divided into small blocks or strips, usually with an electric fence. Each strip is calculated to give enough grazing to the stock for one feed or one day.

It is best to use a *back fence* so that the stock cannot run back over the land they have grazed before. This method makes the best use of land for grazing.

Creep Grazing: Young animals, such as lambs or calves, are allowed on to clean, new pastures through special fences or gates which prevent full grown animals passing through; thus they get the best feed, free from contamination.

Zero Grazing: Stock are not kept on the land at all, but in buildings or yards. The grass is cut and taken to them as required daily.

CULTIVATIONS FOR GRASSLAND

If grassland is down for any length of time, some work will need to be done on it to keep it in good order.

Rolling will be needed for a young ley, and is commonly done on any field which is to be cut for hay or silage. This levels the surface and puts any stones out of the way of mower knives.

Harrowing pulls dead stuff from the bottom of grassland, lets air into the surface of the soil and spreads dung. To avoid spreading wet dung about all over the grass, it is best done in dry weather.

Heavier Cultivations, using heavy harrows or disc harrows, are sometimes done on old grass which has become very matted and dead below.

Topping is done with a mower or a forage harvester to tidy up rough grassland and to prevent grass from forming stem and seed (see page 132). It also controls some weeds, and may be necessary with a young ley which is coming up very weedy.

Spraying is commonly used for weed control on all types of grassland. Among the worst weeds are thistles, docks, ragwort, buttercup and horsetail (found in wet land).

GRASS FOR WINTER USE

It would be ideal if grass grew the whole year round. In our climate, it does not. In order to feed livestock as cheaply as possible through the autumn and winter, grass must be preserved in some way.

There are several methods:

Foggage. Some grass fields can be rested from late August onwards, given a dressing of N, and used for grazing during the period October/November/December.

Types of ley which can be used for this purpose are Italian ryegrass, cocksfoot, and timothy/meadow fescue. Lucerne leys when grown in rows are also used for late grazing.

Silage. This means preserving green crops for winter use in wet form, by a sort of pickling process. The green material—which may be grass, maize or pea haulm—is cut, stacked or put in a container, allowed to settle down tight, covered against the weather, and used when it is needed.

It may be cut and carted straight away, or cut and allowed to wilt for a time—which reduces the amount of water in it. Cutting is done with a mower or with a forage harvester.

According to the way it is cut and treated later, silage is either long (as cut with a mower) or chopped in some way (as with a forage harvester).

Short, young green material (high in protein) makes the best quality silage, but has to be treated with care. It is often necessary to add certain chemicals (known as additives) sold for this purpose.

To make good quality silage, you have to start with good quality grass (or other green stuff). Good silage is light green rather than brown, smells acid (like pickles) rather than too sweet (like tobacco) and is made of good leafy material.

Hay: Grass, and other types of green crop, are cut and dried naturally in the field to make hay.

Most of the feeding value of grass is in the leaf. If grass is cut too late, it will be stemmy and of low feeding value. It should be cut for hay-making at a time which gives a good feeding value and also a reasonable yield. A good time is just before most of the grasses are flowering. May/June is the best time, rather than June/July.

Hay-making in the past was often a long and anxious time. The aim today is quick hay-making—using 'conditioning machines' after the grass is cut; such as tedders, which shake the grass, and which speed up drying in the field.

Hay is baled for storage, and the moisture content should be down to 20 per cent before it is stacked.

In the wetter parts of the country it is more difficult to make hay safely. Partly dried grass can be put on tripods or fences, or put up in small cocks, and allowed to dry naturally. Barn hay drying means partly drying the grass in the field, and then putting it—either loose or baled—into buildings or stacks where air can be blown through it, to dry it thoroughly. Good hay is green, smells sweet and not musty, contains plenty of leaf and not too much stem or flowering heads. Look at it closely, handle it and smell it to test its quality.

Grass Drying produces the highest quality form of preserved grass. It is an expensive process, and usually organized on a factory scale, and so is not found as part of ordinary farming.

Most grass drying is done on a large scale, and dried grass and lucerne are sold as concentrated feeding stuffs.

PRODUCTION OF PRESERVED GRASS

One tonne (1,000 kg) of fresh grass will produce either:

Silage	750 kg
or	
Hay	250 kg
or	
Dried grass	200 kg

Yields

	Average	*Good Yield*	*Our Farm*
Tonnes of hay per acre	1·5	2·0	
Tonnes of hay per hectare	4·0	6·0	

Where hay or silage has to be stored in buildings, stacks, or silos, the figures shown on page 138 for crop storage can help in calculating the space needed.

THINGS TO DO

1. Walk over, inspect, and check the condition of rough grazing, old permanent pasture, and a one-year ley.
2. Look at, handle, and see in both the early growth stage and the heading stage, these plants: Ryegrass, timothy, cocksfoot, agrostis, Yorkshire fog, white clover and lucerne.
3. Watch livestock grazing, see how they actually feed, and study the condition of grassland after they have grazed it—cattle, sheep and horses.
4. Study types of fencing and any other methods of controlling grazing in the field.
5. See, handle, smell and judge the quality of samples of hay and silage—good, average, and poor quality if possible.

QUESTIONS

1. What percentage of grassland is grown on your farm, or on a typical local farm? What types of grassland are there on your farm?
2. How would you tell the difference between ryegrass and agrostis, and between white clover and red clover?
3. What are four important weeds of grassland, and what harm or trouble do they cause?
4. What is an average yield of hay, and an average price per tonne in your district at the present time?
5. How do you judge the quality of hay—what different points do you test?

Appendices

REFERENCE BOOKS

The reader may need to refer to other books for further information or for more detail. The following books are suggested:

Farm Livestock, *Farm Machinery*, *Farm Workshop*, in this Farming Book series.

Farm Management Pocketbook by John Nix, published each year by Wye College, Ashford, Kent.

Soil Management by Davies, Eagle and Finney, published by Farming Press Ltd.

CROP STORAGE

Some crops need to be stored after harvesting. Sometimes, types and varieties need to be kept separate, for marketing or other reasons—e.g. milling and feeding wheats. In order to work out the space needed, and relate this to the capacity available in bins, buildings, silos, or field clamps—use the method shown below.

This will also serve as a means of calculating the amount of a crop in store. Obviously it is not so accurate as weighing. In the end you will only get paid for what is actually delivered to the buyer. What is put into store does not always come out the same. There may be losses due to heating, shrinking, leaking away through holes and cracks, damage by pests and vermin, etc.

1. Measure the store, using the same units throughout.
2. Calculate the cubic capacity, using the right formula according to the shape of the store:

   Square or rectangular $\text{Length} \times \text{Width} \times \text{Height}$

   Triangular clamp or heap $\dfrac{\text{Length} \times \text{Width} \times \text{Height}}{2}$

   Cylinder (e.g. bin or silo) $\text{Radius}^2 \times \text{Height} \times 3{\cdot}142$.
3. Divide this figure according to the material stored, as shown in table at top of page 139. This will give you a figure in tonnes.
4. If you know the number of tonnes of material you need to store, you can calculate the cubic capacity needed by multiplying the number of tonnes by one of the figures. This will give you a figure in cubic metres.

CROP YIELDS

In each section of this book, under the different crops, figures are given showing average crop yields. These are based on Ministry of Agriculture

Crop Storage

Material	*Cubic Metres per Tonne*	*Cubic Feet per Ton*
Wheat, maize	1·3	46
Barley	1·4	51
Oats	1·9	70
Rye	1·4	51
Beans	1·2	43
Peas	1·3	46
Oil seed rape	1·4	50
Potatoes	1·56	56
Sugar beet, mangels, swedes	1·78	64
Silage	1·25	54
Farmyard manure	1·12	40
Baled hay	6·0	8 cubic yards
Baled wheat straw	13·0	15·5 cubic yards
Baled barley straw	11·5	20 cubic yards

figures, prepared each year (and over longer periods) for the whole country. In each case, another figure is given for 'a good yield', showing what a good grower can average in a good farming district. The third space, left blank, is for your own average figures, based on yields in any one year (or averaged out over a longer period).

Yields vary considerably; they tend to increase over the years as varieties and growing methods improve. It is sometimes instructive to compare current yields with those of some years ago. You may find it useful to fill in the table below, so that your references for average yields are up to date.

Crop Yields

Crop	*Tonnes per Acre*	*Tonnes per Hectare*
Wheat		
Barley		
Oats		
Rye		
Beans		
Peas		
Oil seed rape		
Potatoes		
Sugar beet		
Hay		
Wheat straw		
Barley straw		

CROP CALENDAR

Crop Information				*Field Work Done*	*Materials*	*Date*
Crop						
Variety						
Previous crop		Next crop				
Seed rate						
Type of seed						
Sowing method						
Fertilisers applied	N	P	K			
Other plant foods						
Sprays applied						
Yields	average	good	actual			
Storage						
Market						

Copies of this form can be used to record the details of any crop that you study in the field. It also has a use for revision; to test your knowledge, fill in a form for any crop, and check the details afterwards, point by point.

METRIC CONVERSION TABLES

BRITISH TO METRIC	METRIC TO BRITISH

LENGTH

British to metric	Metric to British
1 inch (in) = 2·54 cm or 25·4 mm	1 millimetre (mm) = 0·0394 in
1 foot (ft) = 0·30 m	1 centimetre (cm) = 0·394 in
1 yard (yd) = 0·91 m	1 metre (m) = 1·09 yd
1 mile = 1·61 km	1 kilometre (km) = 0·621 miles

Conversion Factors

British to metric	Metric to British
inches to cm × 2·54	centimetres to in × 0·394
or mm × 25·4	millimetres to in × 0·0394
feet to m × 0·305	metres to ft × 3·29
yards to m × 0·914	metres to yd × 1·09
miles to km × 1·61	kilometres to miles × 0·621

AREA

British to metric	Metric to British
1 sq. inch (in^2) = 6·45 cm^2	1 sq. centimetre (cm^2) = 0·16 in^2
1 sq. foot (ft^2) = 0·093 m^2	1 sq. metre (m^2) = 1·20 yd^2
1 sq. yard (yd^2) = 0·836 m^2	1 sq. metre (m^2) = 10·8 ft^2
1 acre (ac) = 4047 m^2 or 0·405 ha	1 hectare (ha) = 2·47 ac

Conversion Factors

British to metric	Metric to British
sq. feet to m^2 × 0·093	sq. metres to ft^2 × 10·8
sq. yards to m^2 × 0·836	sq. metres to yd^2 × 1·20
acres to ha × 0·405	hectares to ac × 2·47

VOLUME (LIQUID)

British to metric	Metric to British
1 fluid ounce (1 fl oz) 0·05 pint) = 28·4 ml	100 millilitres (ml or cc) = 0·176 pints
1 pint = 0·568 litres	1 litre = 1·76 pints
1 gallon (gal) = 4·55 litres	1 kilolitre (1000 litres) = 220 gal

Conversion Factors

British to metric	Metric to British
Pints to litres × 0·568	litres to pints × 1·76
gallons to litres × 4·55	litres to gallons × 0·220

WEIGHT

British to metric	Metric to British
1 ounce (oz) = 28·3 g	1 gram (g) = 0·053 oz
1 pound (lb) = 454 g or 0·454 kg	100 grams = 3·53 oz
1 hundredweight (cwt) = 50·8 kg	1 kilogram (kg) = 2·20 lb
1 ton = 1016 kg or 1·016 t	1 tonne (t) = 2204 lb or 0·984 ton

Conversion Factors

British to metric	Metric to British
ounces to g × 28·3	grams to oz × 0·0353
pounds to g × 454	grams to lb × 0·00220
pounds to kg × 0·454	kilograms to lb × 2·20
hundredweights to kg × 50·8	kilograms to cwt × 0·020
hundredweights to t × 0·0508	tonnes to tons × 0·984
tons to kg × 1016·0	
tons to tonnes × 1·016	

INDEX